Pascal Applications for the Sciences

Pascal Applications for the Sciences

Richard E. Crandall

Department of Physics
Reed College
Portland, Oregon

A Wiley Press Book

JOHN WILEY & SONS, INC.
New York · Chichester · Brisbane · Toronto · Singapore

Publisher: Judy V. Wilson
Editor: Theron Shreve
Composition and Make-up: Cobb/Dunlop, Inc.

Library of Congress Cataloging in Publication Data
Crandall, Richard E., 1947–
   PASCAL applications for the sciences.

   (Wiley self-teaching guides)
   Includes index.
      1. Science—Data processing—Programmed instruction.
   2. PASCAL (Computer program languages)—Programmed
   instruction.   3. Programming (Electronic computers)—
   Programmed instruction.   I. Title.
   Q183.9.C73   1984   001.64'24   82-24832
   ISBN 0-471-87242-3

Printed in the United States of America

84 85 10 9 8 7 6 5 4 3 2 1

Preface

Pascal Applications for the Sciences has evolved as a result of a common dilemma facing teachers of science: How can a diverse group of students be taught scientific programming in a minimum amount of time? In 1978, the Reed College Physics Department encountered such a dilemma when they decided to adopt Pascal as the language of choice for physics courses. Even at the introductory level, students have diverse approaches to science and may be committed to a specific field such as biology, physics, medicine, or chemistry. The teachers had to resolve two questions: How can these students be brought to a level of scientific programming that is adequate for basic applications in their chosen disciplines? How can this be done without taking too much time away from the normal course of physics study? The dilemma was compounded by the fact that, even though some marvelous student assistants lent their talents to the Pascal sessions, the average introductory student had to spend considerable time alone at a terminal, in self-study mode.

This book incorporates what I have learned over a 5-year period about how students of scientific programming prefer to pace themselves. The course materials for the Pascal sessions at Reed College have been refined, tested, and amplified to provide a means by which students can acquire scientific programming skill in a short time. The book is for scientists and students of science who wish to achieve the expertise needed to solve problems in their particular disciplines. This book is not a traditional computer science book—it is a guide for any scientist. It contains a review of Pascal structure, but it focuses on applications. For this reason, I recommend that students use a standard Pascal text for details on Pascal syntax and structure. In the original college sessions, the materials in this book were used together with a standard Pascal text. This combination proved so effective that serious students were able to achieve good levels of expertise in one school quarter.

Pascal Applications for the Sciences has the following sequence: Pascal review, mathematical programming, equation solving, statistics, and graphics. The book then branches into four advanced chapters, with applications for mathematics, chemistry, physics, and biology. All the chapters include exercises that cover typical applications, as well as exercises of an exploratory nature.

The five appendixes contain scientific libraries as Pascal include files. These libraries have computer packages students will find useful in their studies; they cover graphics, matrices, statistics, special functions, and dynamical models.

Throughout the book the sequence of text—exercises—text—exercises—and so on is followed. I have found that there is a limit to how long a programmer can read text material before an exercise series is appropriate. The length of each text segment has been designed accordingly. For those facets of study I take to be more important, there are noticeably more exercises following.

Above all, *Pascal Applications for the Sciences* is designed to provide you with efficient, unified training. Scientists of today have so much to learn in their chosen areas that the acquisition of programming skill should not usurp an inordinate amount of time away from traditional scientific pursuit. You will have been successful with this book if you can achieve a certain independence from it. It is my hope that you will eventually be able to address any problem in your field, using the book only in matters of reference.

I would like to thank the introductory students whose feedback over the years has been indispensable for this project. I also wish to thank the instructors R. Henley, R. Whitnell, S. Swanson, W. Wood, and T. Abbott whose excellent works are included in this book. I am indebted to faculty members D. Hoffman, R. Reynolds, J. Delord, R. Bettega, N. Wheeler, D. Griffiths, J. Buhler, R. Mayer, T. Dunne, and J. Dudman for programs, corrections, and general support of the project. I thank computer colleagues C. Green, G. Schlickeiser, E. Roberts, G. Ross, M. Penk, and D. Basin for their gracious offerings of expertise. I especially thank S. Stearns for allowing me to transfer his insightful course materials for Pascal as it pertains to biology. I am grateful to Intel Corporation, Hewlett-Packard, Tektronix, and Apple Computer personnel for support in the form of equipment, encouragement, and general interest in the project. I am indebted to A. Marcus, M. Blair, R. Kilgore, R. Raber, M. Lindquist, and C. Delord for their inestimable aid in generation of the manuscript.

Richard E. Crandall
Portland, Oregon

How to Use This Book

Pascal Applications for the Sciences is designed to lead you from a novice level of scientific programming to a level of expertise at which you can routinely solve problems in your field of science. If you need a review of Pascal, begin with Chapter 1. If you already have some Pascal skills, look over the exercises in Chapter 1 to make sure that you can do them, then move on to Chapter 2 where applications begin. If you are already an experienced scientific programmer, try the more difficult exercises in Chapters 2 to 5 and the material in the advanced Chapters 6 to 9.

Each chapter has the following structure:

Text
Exercises
Text
Exercises
. . . .

Throughout the book the theme is "explore!" and in this spirit many of the exercises, notably the final ones of each exercise section, have been included as challenges for your creative abilities. There are references at the end of the book for most chapters so that you may pursue the scientific concepts discussed in the text.

You will notice that instructions in Chapters 6 to 9 are less direct and that the burden of creativity is passed to you. Many of the exercises in these chapters have the flavor of projects as opposed to test questions.

Good luck with Pascal and with science!

Contents

1 | Pascal Review

These machines have no common sense; they have not yet learned to "think," and they do exactly as they are told, no more and no less. This fact is the hardest to grasp when one first tries to use a computer.

Donald E. Knuth
The Art of Computer Programming

GETTING STARTED

Pascal can be used in a great variety of ways such as to display data, to analyze data, to verify theoretical predictions, and to suggest new lines of scientific thought. It is suitable for applied as well as theoretical science. Many books only cover the abstract features of Pascal, such as detailed syntax rules and algorithm structure. This book, however, emphasizes the *utility* of the language. In order to benefit from the examples and exercises, it is important that you first obtain a working knowledge of Pascal. Since this book teaches you to instruct your computer to perform a wide variety of tasks, you must learn how to tell your computer exactly what you want it to do.

Exercise

Learn how to edit programs on your computer system. Compile these programs and arrange your files. Your own system's documents are best for this. Use a standard Pascal text, such as the ones listed in Chapter 1 references, for matters of detailed syntax. It is a good idea to read this chapter with a standard text at your side. If you can successfully do the exercises in this chapter, you are ready to move to Chapter 2 where essential mathematical applications begin.

ELEMENTARY SYNTAX

Most Pascal references contain *syntax diagrams*, which are roadmaps of the overall program and show graphically what structures are legal. Some programmers use references that have all relevant Pascal syntax diagrams; some never use them. In any case, they are useful in the preliminary stages of understanding. This chapter presents some diagrams that frequently occur in application programs.

The most important diagram is for the *program* itself (Figure 1.1). The syntax diagram in Figure 1.1 shows how to start editing a Pascal program. You follow the diagram from left to right and always start with the word "program."

1

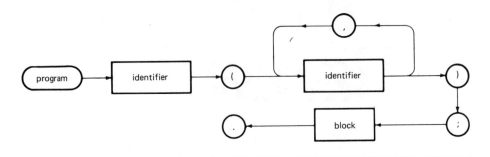

Figure 1.1

The word "program," enclosed in an oval box, is a *keyword,* or *reserved word,* in Pascal. The circled items are *operators,* which are Pascal's punctuation symbols. Items in square boxes are defined by other syntax diagrams. As you will see, what is normally thought of as a programming task will involve almost exclusively a structure called the *block.*

Identifiers are Pascal's words for labeling various objects. For example, the first identifier on the left in Figure 1.1 will be the name of the program. The identifiers following it in parentheses are called *file identifiers.* Generally, these take the form of some combination of the two words "input" and "output".

An identifier has its own roadmap. The diagram in Figure 1.2, for example, means that an identifier is any sequence of letters and digits beginning with a letter. Thus "x", "x5", "voltage", and "position" are legitimate Pascal identifiers, whereas "3x" and "x.3" are not. Verify for yourself that this is the case by trying to trace each of these six identifiers through the syntax diagram in Figure 1.2.

The program section labeled block is essentially all of the Pascal program. The section preceding the block is often called the *program header.* A block has the structure shown in Figure 1.3.

Examples of statements are shown in Figure 1.4. Each line of Figure 1.4 is a separate statement, as indicated in the abstract syntax diagram in Figure 1.3, where semicolons separate all statements.

Since the Pascal language provides us with an enormous multiplicity of possible statements, we will discuss a specific statement and its syntax as it becomes useful.

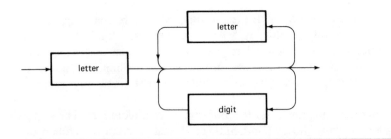

Figure 1.2

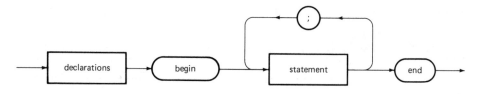

Figure 1.3

```
x:= 15;
while (y = 0) do search;
if (not tadpole) and (alive) then frog:= true;
if angle(x) = 1 then z:=5 else writeln('help!');
clearscreen;
z:= trunc(x+0.5) - y mod p + m*n;
```

Figure 1.4

```
program newton(output);
begin
    writeln('f = ma')
end.
```

Figure 1.5

Figure 1.5 is a legitimate Pascal program. Notice that the program name, an identifier, is "newton", that the file identifier is "output", that there is no declaration section, and that the single statement in the program is

writeln('f = ma')

This statement is a *procedure call*. It activates the standard procedure "writeln" (for write line), which writes things out. The items inside the parentheses, the *arguments* to "writeln", are what will be printed. In this case, the argument is a *string constant*, a group of characters enclosed in single quotation marks. Such quoted characters are taken literally. Therefore, when compiled and executed, this program will print

f = ma

and that is all it will do.

Exercises

1. By using the syntax diagram for identifiers, indicate which of the following are valid identifiers:

loopcount	xy	2x	2 + A
loop count	x y	x2	A − 2

```
program gaslaw(input, output);
begin
    writeln('ideal gas law:');
    writeln('PV = NRT')
end.
```

Figure 1.6

2. Consider the program shown in Figure 1.6. By using syntax diagrams, list these items:
   - (a)   The first keyword in the program
   - (b)   The first identifier (this will be the program name)
   - (c)   The file identifiers
   - (d)   The whole block
   - (e)   The operator that separates statements
   - (f)   The identifier naming the one procedure used in this program
   - (g)   The string constant representing the first piece of output printed

Answers

1. Valid identifiers: loopcount, xy, x2. None of the others can be properly traced through the syntax diagram.
2. (a)   program
   - (b)   gaslaw
   - (c)   input, output
   - (d)   begin
     - writeln('ideal gas law:');
     - writeln('PV = NRT')

     end
   - (e)   ;
   - (f)   writeln
   - (g)   'ideal gas law:'

WRITING OUTPUT

Consider the program shown in Figure 1.7. When run, this program will print the following:

newton's second law:
f = ma.

This example contains two items of interest. First, note that two apostrophes are used in a string constant when you want a single one to appear on output; second, after "writeln" has printed all its arguments, it prints a *carriage return/line feed* that forces the next item printed to appear on a new line. Pascal also has the

```
program newtonl(output);
begin
     writeln('newton''s second law:');
     writeln(' f = ma.')
end.
```

Figure 1.7

```
program manylines(output);
(* Quote-printing program *)
begin
    write('every body continues', ' in its state  ');
    writeln('of rest');
    write('or in uniform motion in a right line unless');
    writeln;
    writeln('it is compelled to change', ' that state by forces');
    writeln('impressed upon it. ');
    writeln;
    writeln('                       -- newton')
end. { manylines }
```

Figure 1.8

procedure "write", which allows the data printed by several different statements to appear on the same line.

Both "write" and "writeln" can take several arguments, and both will print all the arguments given them on the same line. Further, "writeln" can be used without any arguments. The program in Figure 1.8 will print the following six lines of text:

```
every body continues in its state of rest
or in uniform motion in a right line unless
it is compelled to change that state by forces
impressed upon it.
             — —newton
```

The program includes a *comment* that explains the program's purpose. Comments are ignored by your machine when a program is compiled, but they help to explain a program's structure and function to the reader. Another reason to include comments is that when you return later to edit a section of your program, you may have forgotten your original rationale. This predicament is easily averted by way of good commenting habits. Comments are usually enclosed in symbols (* *) or in "braces" { }.

Exercises

1. Consider the program in Figure 1.9. How many lines of output will this program produce? What is the program name? For what phrase do you think the program name is an abbreviation?

```
program glop(input, output);

begin
    write('e=ir');
    write(' ');
    writeln('e=mc squared');
    writeln('div b = 0     epsi = hpsi')
end.
```

Figure 1.9

2. Write your own program to print the following text:

    the test of
    all
    knowledge is
    experiment!

    —feynman

    One blank line should be printed before the author's name. Enter your program on the computer, compile it, and run it.

Answers

1. There are two lines of output. The program name is "glop", which stands for great laws of physics. The great laws are Ohm's law, Einstein's equation, a Maxwell equation, and the Schrödinger equation.
2. The last three lines of such a program should be

    ```
    writeln;
    writeln('—feynman');
    end.
    ```

IDENTIFIERS AND DECLARATIONS

There are two kinds of identifiers. *Standard identifiers* are automatically provided by Pascal. For example, "writeln" is a standard identifier that refers to a procedure that writes things. The other kinds of identifiers are invented by the user. You must declare them before they are used so that Pascal will know what each identifier names. This is done in the *declarations section* of a program, which has the form shown in Figure 1.10.

   The simplest thing to identify is a constant. *Constants* are numbers, such as 5 or 3.1415926535, or strings, such as 'this string', which do not change. We can attach identifiers to constants with a *constant definition*. For example, we could define the identifier "e" to have the value 2.718281828. The constant declaration section of a program has the structure shown in Figure 1.11. The program "newton2" in Figure 1.12 has an effect identical to that of the program "newton" given previously.

One of the most useful things that can be identified is a variable. *Variables*, like constants, represent numbers, characters, and so on. Unlike constants, however, the *value* of a variable can be changed.

Imagine a variable as a box in which you place items to look at later. This box could be said to have a very specific shape, which is its type. The *type* determines what variables you can and cannot put into the box. For example, you can put the numbers 1, −3, 7/8 in variables declared to hold the type called real numbers.

All variables must be declared as such and assigned a specific type. This is done in the *variable declaration section* of a program illustrated in Figure 1.13. We will mainly be concerned with the two numerical types *integer* and *real*.

You can also declare your own customized types, procedures, and functions. These options come up later in the book.

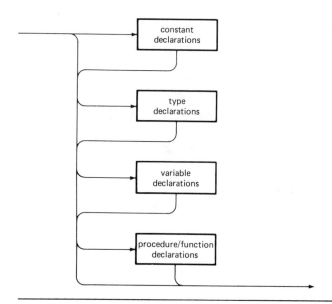

Figure 1.10

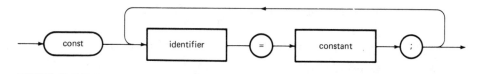

Figure 1.11

```
program newton2(output);
const
      law2 = 'f = ma ';
begin
      writeln(law2);
end.
```

Figure 1.12

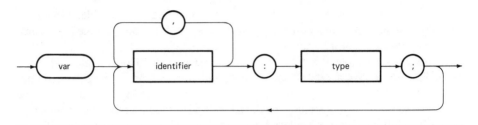

Figure 1.13

Exercises

1. Indicate which of the Pascal *types*—"integer", "real", "boolean", "char", or "illegal"—each of these values are:

   1.0 −99.7 6x true false maybe 'h' '65'
   1−0.2 99.012 −678

   For this and other problems you may wish to consult a Pascal reference (see Chapter 1 references).

2. For the program in Figure 1.14, list the following:
   (a) All standard identifiers
   (b) All keywords
   (c) All user-defined identifiers

Answers

1. 1.0   real
   −99.7   real
   6x   illegal

```pascal
program ex1(input, output);

(* print the area and circumference of a circle. *)
const
    pi = 3.14159265358979;
var
    radius, area: real;
begin
    radius := 2.35; (* set the variable 'radius' to a number. *)
    area := radius * radius * pi; (* calculate the area. *)
    writeln('radius =', radius);
    writeln('area =', area);
    writeln('circumference =', 2 * pi * radius)
end. { ex1 }
```

Figure 1.14

true	boolean
false	boolean
maybe	illegal
'h'	char
'65'	illegal
1−0.2	real
99.012	real
−678	integer

2. (a) Input, output, real, and writeln
    (b) Program, const, var, begin, and end
    (c) Exl, pi, radius, and area

CALCULATIONS IN PASCAL

In order to use variables, you must give them values. This is often done with an *assignment statement,* as shown in Figure 1.15. The section labeled "expression" is where you actually do calculations. The value calculated is then stored in the variable identified to the left of the ":=" operator. Thus,

x := 2

sets the variable "x" to the value 2.

Expressions can be extremely complicated. Consider the examples in Figure 1.16. Two things are immediately apparent. First, expressions use *operators* in

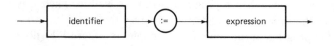

Figure 1.15

Expression	Value	Explanation
15 + 2	17	----
15 * 2	30	----
15 / 2	7.5	real number division
15 div 2	7	integer division
15 mod 2	1	remainder after division
15 * 2 + 1	31	----
15 * (2 + 1)	45	----
15 / 1 + 1	16	same as (15 / 1) + 1
15 / (1 + 1)	7.5	----

Figure 1.16

normal algebraic fashion. Second, types are relevant for some operations, particularly division. Basically, division with the operator "div" *truncates* the result so that it is an integer. Division with the operator "/" forces the result of the division to be real.

If you have variables that have values, you can use them in expressions. For example, the program segment

```
x := 5;
y := 7;
mean := (x + y) / 2;
```

will set the variable identified by "mean" to the value 6.0.

In addition to operators, Pascal provides you with *functions* to use in expressions. For example, "sqrt" identifies a function that takes the square root of a number, e.g., "sqrt(9)" returns the value 3.0. The numbers passed to functions are called *arguments,* such as those given to procedures. These arguments may themselves be expressions, for example,

```
5 * sqrt(1 + x * y)
```

There are many functions available in Pascal. Important ones are "sqr", which squares numbers; "ln", which returns the value of the natural logarithm of its argument; "exp", which returns e^x; "sin" and "cos", which take arguments in radians; and "arctan", which yields the inverse tangent of its argument. Thus,

```
x := 1/sqr(sin(3.1415926536/4))
```

will give the variable "x" the value 2.0 because $\sin \pi/4 = 1/\sqrt{2}$.

Variables can be given values when a program is run, as well as when it is written. This is done with the procedure "read", which gives the variables those values that are input to the program at run time. For example, in Figure 1.17 note the use of "read" and of an expression as an argument to "writeln". Also, realize that the file identifier "input" *must* be present in the program header of any program that uses "read".

```
program avg(input, output);
(* get three numbers and print their average. *)
var
    a, b, c: real;
begin
    read(a, b, c);
    writeln((a + b + c) / 3)
end. { avg }
```

Figure 1.17

Exercises

1. Determine the Pascal *type*, "real", "integer", or "char", for each of the following expressions:

   sqrt(7)   trunc(7.6665)   sin($\emptyset$)
   ord('9')   chr(9)   (x mod a)/1.2
   7 div 5

2. Write a program that will read in a real number, store it in a variable "x", then print the natural logarithm, ln(x). Compile this program, and run it for x = 0.5, 1, 52, and −1. Note the *run-time error message* that is printed when x = −1.

3. A block of mass m is being whirled around in a circle on a string of length r. What is the tension on the string? If w is the angular velocity of the block, then the tension t is given by

   $$t = ma = mrw^2$$

   Write a program that reads the mass of the block, the length of the string, the angular velocity, and then prints the tension. You will want to have a statement similar to

   writeln('tension =',t);

   in your program. Verify that if m = 0.5 kilograms, r = 1 meter, and w = 5 radians per second, then t = 12.5 newtons.
     If "Tmax", the greatest allowed tension for your string, is 500 newtons and the string is 1 meter long, how fast can you whirl a 1-kilogram block without breaking the string? Use your program and a trial-and-error method to determine this value of w.

4. Modify the program that you wrote for exercise 3 as follows. Change the "writeln" statement so that it reads

   writeln('tension =',t:7:3);

   Note the effect that this change has on your program's output.

Answers

1. sqrt(7)   real
   trunc(7.6665)   integer
   sin(0)   real
   ord('9')   integer
   chr(9)   char
   (x mod a)/1.2   real
   7 div 5   integer

2. The real variable x can be read in with the statement

   read(x);

   and its logarithm can be printed with the statement

   writeln(ln(x));

3. The line that computes tension should be

   t := m * r * w * w;

   The correct maximum value for w is 22.36.

4. Formatting is a good idea when you do not need full precision or when you need clean tabulation of output. For tension 12.5, for example, the :7:3 formatting should give you 12.500 as the answer.

2 | Mathematical Programming

The objects have, and can have, no properties . . . . Indeed an object cannot be known or conceived otherwise than as a complex of [relations to other things]. In mathematical phrase: things and their properties are known only as functions of other things.

J. B. Stallo
The Concepts and Theories of Modern Physics (1960)

MOTIVATION

Stallo's words help to explain why mathematics is ubiquitous in the sciences: We model relations between things with relations between quantities. Chapter 1 reviewed how to input and output quantities into and out from programs. The next step is to learn how to interrelate concepts within the program block. Pascal will not be useful until you can perform complex mathematics within your programs. This chapter is therefore concerned with various techniques that will enable you to begin modeling phenomena with mathematics, using Pascal.

Exercise

Review the concepts of sequences, summations, and limits by consulting the references for Chapter 2, specifically Davidson and Marion (1972). In many cases this chapter will enhance your understanding of these ideas. As you use the computer to manipulate quantities according to your instructions, you will produce many abstract phenomena discussed in the references.

SEQUENCES

Consider this statement from a biology text: "The biological process that leads most naturally to difference equations is population growth" (Stearns, 1980). Difference equations will be discussed later. For the moment the idea is that they generate *sequences*. Consider, for example, the problem of rabbit reproduction based on these axioms:

1. We start with one rabbit in year 1.
2. All rabbits live forever.
3. A given rabbit reproduces first at age 2 and then every year thereafter.
4. A reproduction year for a rabbit involves exactly one new birth.

If you construct a table of number of rabbits f_n in year n, you find the following:

Number of rabbits f_n	Year n
1	1
1	2
2	3
3	4
5	5
8	6
13	7
.	.
.	.
.	.

The numbers f_n can be thought of as generated by the *difference equation*:

$$f_{n+1} - f_n = f_{n-1}$$

which shows that last year's rabbits f_{n-1} is the difference between next year's rabbits and this year's rabbits. Inspect the table to satisfy yourself that this assertion is true. This equation can also be put in the form of a *recurrence relation*

$$f_{n+1} = f_n + f_{n-1}$$

to show how next year's rabbits f_{n+1} is the sum of this year's rabbits and last year's rabbits. Although our example is relatively simple, difference equations, or their associated recurrence relations, are sometimes quite complicated. The numbers generated in the rabbit problem are the well-known Fibonacci numbers (Roberts, 1977). *Fibonacci numbers* are an infinite series of which the first two terms are 0 and 1 and each succeeding term is the sum of the two preceding terms.

Let us write a program to print out every Fibonacci number that is less than some predetermined number. There will not be too many Fibonacci numbers less than your given number, because the f_n grow so fast. In our rabbit example, in 500 years f_n will have more than 100 decimal digits. In order to write the program we shall use a loop. A *loop* is a program section that executes repetitively, changing certain internal variables at each execution pass. Pascal has several options for such repetitive execution. We will first use the *repeat-until statement,* whose syntax diagram is shown in Figure 2.1. The looping process, sometimes called *iteration*, occurs until the boolean expression after "until" becomes true.

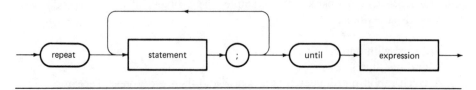

Figure 2.1

```
program fibs(output);
const maxf = 1000;
var    past, present, future: integer;
begin
    past := 0;
    present := 1;
    (* loop, printing numbers begining with f(1). *)
    repeat
        future := past + present;
        writeln(future);
        past := present;
        present := future;
    until present >= maxf;
end.
```

Figure 2.2

We shall need variables to hold some of the f_n at any given time. Let us call these "past", "present", and "future". Because the number of rabbits is an integer, these variables may be of type integer. We also need a predetermined maximum number, which we will call "maxf"; it will appear in the "until" statement. Now consider the working program in Figure 2.2. Note the "repeat-until" loop is broken only when the condition "future = maxf" is satisfied, so that sometimes we may get a printout of an extra Fibonacci number. *Question:* How can we avoid getting any Fibonacci numbers that meet or exceed the value of "maxf"? *Answer:* There are many ways to do this. We can "writeln" something besides "future", using a different kind of iterative loop or change the loop condition in the "until" statement (see the following exercises).

Recursion relations occur in many scientific settings. Population biology has naturally occurring relations because of the way that future populations are determined by the present one. In physics, such processes are usually described by *differential equations,* which in a certain sense are difference equations with infinitesimal (as opposed to integer) differences. On a computer, you almost always solve differential equations using recurrence relations.

You should be aware that Pascal has the capability for *recursive procedures* and *recursive functions.* These methods can be used to compute sequences and are considered in Chapter 6.

Exercises

1. Explain why changing the constant "maxf" to 144 in the program "fibs" will result in the printing of 144. Change the program so that for any maxf value, the printout will be only those Fibonacci numbers that are less than maxf. This task is very common and is important to understand.
2. Explain why the lines "past := present" and "present := future" cannot be interchanged. State a general rule for the order in which such lines must appear in your program.
3. Write a program to generate the numbers P_n where $P_0 = 1$, $P_1 = 3$, and the recurrence relation is

    $$P_{n+1} = 2 * P_n + P_{n-1} \qquad n = 0, 1, \ldots$$

Use this program to calculate P_{20}. Save this value for exercise 5. (How? You can save it on a disc, print it out, or write it down on a piece of paper when it comes up on your terminal.) On your system you may have to define P_n as real so that it can grow sufficiently large.

4. Modify the program you wrote in exercise 3 to calculate the numbers Q_n satisfying the *same* recurrence relation but with initial values $Q_0 = 1$, $Q_1 = 2$. Find Q_{20} and save it for exercise 5.

5. Write a short program to print out the real number P_{20}/Q_{20} and also its square.

6. You have computed the square root of 2 with the techniques in exercises 3 to 5. The reason it works is that one way to express the square root of 2 is

$$\sqrt{2} = 1 + \cfrac{1}{2 + \cfrac{1}{2 + \cfrac{1}{2 + \cdots}}}$$

Now write just one program that combines those in exercises 3 to 5. Use it to get the number defined by the continued fraction in this exercise but with the 2's on the right-hand side replaced by 3's. The result should be the number $(1/2) * (\sqrt{13} - 1)$.

7. Suppose that a rabbit is incompressible and occupies a volume of 0.01 cubic meters. Realizing that the radius of the earth is 6400 kilometers and that the volume of a sphere is $(4/3) * pi * r^3$ for radius r, write a program that will print out how many years will elapse before one rabbit and her accumulated descendants occupy a volume equal to that of the earth.

*8. Denote the continued fraction for $\sqrt{2}$ by $(1,2,2,2,\ldots)$ as in exercise 6. Compute the natural base $e = 2.718281828\ldots$ by using the fraction $e = (2,1,2,1,1,4,1,1,6,1,1,8,\ldots)$.

*9. Extend the program for exercise 7 to print out how many years must elapse for the velocity of expansion of the sphere of rabbits to exceed the speed of light (taken to be 1e16 meters per year).

*10. Consider the following recurrence problem. Choose a positive integer n greater than 1. Define the function f(n) to be $3 * n + 1$ if n is odd and n/2 if n is even. Apply the function f to the number n, apply f again to the result, and so on, until you reach the value 1. For example, if the starting number is 7, then f(7) is 22, $f(22) = 11$, and you get the sequence

7, 22, 11, 34, 17, 52, 26, 13, 40, 20, 10, 5, 16, 8, 4, 2, 1

*Exercises marked with an asterisk are special exercises. They are problems of an exploratory nature and generally require deeper knowledge of the underlying scientific or mathematical principles.

Write a program that asks for a starting number and prints out the sequence so generated. *Question:* Will every starting number n give a terminating sequence? *Answer:* Unknown (Crandall, 1978).

Answers

1. The number 144 would be printed because the "until" statement executes after "writeln", and 144 is indeed a Fibonacci number. To avoid printing Fibonacci numbers equal to "maxf" you can do

    if future < maxf then writeln(future);

    or you can change the exit condition of the loop to

    until past + present >= maxf;

2. If you interchanged the two lines, both past and present would be equal (and equal to future) after the lines execute, thus destroying the recurrence relation you are solving. Remember Knuth's words about how the computer does exactly what you tell it to do.
3. The main loop should look similar to the following:

    ```
    repeat
        n := n + 1;
        future := 2 * present + past;
        past := present;
        present := future;
    until n = 20;
    ```

    where you have also set past := 1, present := 3, and n := 1 initially. Then p20 will be the final value of "future" when the loop exits. *Caution:* You may have to declare future, past, and present to be reals instead of integers so that your machine will be able to accommodate the large answer:

    p20 = 54608393

4. The only modification needed is to initialize "present" to 2 instead of 3, giving q20 = 38613965.
5. If p and q are defined as constants, the single line

    writeln(p/q)

    will print p/q. For the p20 and q20 calculated from exercises 3 and 4, respectively, the result is the square root of 2 correct to 14 places.
6. A little thought reveals that for the infinite fraction

    $$1 + 1/(a + 1/(a + (a \cdots$$

    the relevant recurrence relations are

```
Pfuture := a * Ppresent + Ppast;
Qfuture := a * Qpresent + Qpast;
```

with initial values

```
Ppast := 1; Ppresent := a + 1;
Qpast := 1; Qpresent := a;
```

Then the fraction evaluated at any finite place will be the number Pfuture/Qfuture, and this ratio will converge to the desired result as more and more passes of the recurrence loop are performed.

7. Since it takes 100 rabbits to fill a cubic meter and there are

$$(4/3) * pi * r * r * r = 1.0983e4$$

cubic meters in the earth, we will need 1.098e26 rabbits to fill the earth. In that the 127th Fibonacci number is about 1.56e26, we need only wait 127 years for the rabbits to fill the volume of the earth entirely.

SERIES

A typical series having no known closed form is as follows:

$$\sum_{n=1}^{\infty} n^{-3} = \frac{1}{1^3} + \frac{1}{2^3} + \frac{1}{3^3} + \frac{1}{4^3} + \cdots$$

This number is called $\zeta(3)$, read zeta of three. Only numerical approximations have been obtained for this sum. For example, you can work out the sum of the first 10000 summands (i.e., including the term 10000^{-3}). It is possible to find finite numerical sums using these steps:

1. Declare a sum variable called, e.g., "sum". This is often real, as for the example above, but it can be an integer (see the following exercises).
2. Determine the nature and degree of desired approximation. There is more than one criterion for stopping a sum.
3. Execute a loop that repeatedly changes the value of "sum" by the proper amount on each pass. In the example above the value of "sum" would be incremented by 5^{-3} on the fifth pass (if "sum" were set to zero before the first pass).
4. Print out the final value after the loop stops. (Sometimes we wish to print out values while the loop is calculating.)

Often step 2 is the most problematic because it is not always easy to choose the correct criterion for the approximation. Three possible criteria are as follows:

1. The number of terms summed has exceeded some limit.
2. The size of the term just computed is sufficiently miniscule.

3. The value of "sum" is within a desired range of a known or anticipated result.

There are also other criteria for stopping a calculating loop. You might, for example, perform the loop until a certain amount of computer time is used up. This is similar to, but not exactly the same as, criterion 1.

Programs that compute the inverse-cubes sum are given in Figure 2.3. Note the differences among these programs. Whereas the interiors of the "repeat-until"

```
program crit1(output);
const
    maxn = 200;
var
    n: integer;
    sum: real;
begin
    sum := 0;
    n := 1;
    repeat
        sum := sum + 1 / (n * n * n);
        n := n + 1
    until n > maxn;
    writeln(sum)
end.
```

```
program crit2(output);
var
    n: integer;
    sum, error: real;
begin
    sum := 0;
    n := 1;
    repeat
        error := 1 / (n * n * n);
        sum := sum + error;
        n := n + 1
    until error < 1.0e-5;
    writeln(sum)
end.
```

```
program crit3(output);
const
    bookvalue = 1.2020569031596;
var
    n: integer;
    sum: real;
begin
    n := 1;
    sum := 0;
    repeat
        sum := sum + 1 / (n * n * n);
        n := n + 1
    until abs(sum - bookvalue) < 1.0e-6;
    writeln(sum)
end.
```

Figure 2.3

loops are all similar, the declarations and the loop exit conditions (the boolean expressions of the "until" statements) are all different. Which method is best depends on your intention. For example, program "crit3" exits when the sum is sufficiently close to a known theoretical value. On the other hand, program "crit2" exits when the last term summed is sufficiently small, but this does not mean the sum itself is as accurate as you might think. In the problem of finding $\zeta(3)$, for example, the condition of "crit2" that the final term be less than 1.0e−5 only gives the sum itself to about three, not five, places. Still, "crit2" is useful if the size of the smallest term can be theoretically related to the accuracy of the sum computed.

The first example, "crit1", is perhaps the easiest to use because the exit condition is simple and a known amount of computer time will be used.

Note that the program "crit3" uses the function "abs" even though the expression (sum − bookvalue) is always positive. This is to emphasize the importance of preventing programming errors as a result of omission of "abs" in those applications where (sum − bookvalue) may actually change sign during the program run.

Exercises

1. Find the sum $1/(1*2) + 1/(2*3) + 1/(3*4) + \cdots$ with the accuracy criterion that the last term added be less than 1e−4. After you determine the approximate value, guess what the infinite sum is. This is a good example of a series that can be summed in your head faster than on the computer. Do you see how?

2. Find the value of the number $1/1 + 1/2 + 1/3 + \cdots + 1/1000 - \ln(1000)$. (You only subtract the logarithm once.) This result will be very close to *Euler's constant* γ, an interesting number whose value (for 1000 replaced by n, n taken to ∞) is given in many tables.

3. Compute the number $S = 1/1^4 + 1/2^4 + \cdots + 1/200^4$. Then compute the number "sqrt(sqrt(90 * s))" and guess the value of the infinite sum of inverse fourth powers.

4. Find the area under the curve $f(x) = 1/(1 + x^2)$, for $0 < x < 1$, by adding up the areas of 10000 little rectangles packed uniformly from $x = 0$ to $x = 1$. This is an approximation to the integral

$$\int_0^1 f(x) \ dx$$

5. Sum the first 1000 terms of the series

$$\sum_{\substack{n=1 \\ n \ \text{odd}}} \frac{\sin(n*x)}{n}$$

for various values of x, including $x = 0.1, 0.2, 0.3, 1.0, 1.5$. Something peculiar happens because the full infinite sum is a certain kind of wave when plotted as a function of x. Using your numerical results, describe the wave.

6. Show with a program that the sum of the first 52 integer squares is divisible by

7, that is, show (sum) mod $7 = 0$, or alternatively, show that the sum of all terms (sqr(j) mod 7) is zero. What is better about the second method?

Answers

1. The exact result is 1, so the "crit3" approach can be used. Alternatively, you can use "crit2" with "until error $< 1e-4$", but as stated in the text this termination condition requires mathematical knowledge of how the overall accuracy of the sum depends on the size of the last summed term. Here is how to do the sum in your head. Notice that the expression telescopes, i.e.,

$$1/(1 * 2) = 1 - 1/2$$
$$1/(2 * 3) = 1/2 - 1/3$$
$$1/(n * (n + 1)) = 1/n - 1/(n + 1), \text{ etc.}$$

so that the series is equivalent to

$$(1 - 1/2) + (1/2 - 1/3) + (1/3 - 1/4) + \cdots$$

and all that remains as we cancel terms is the 1.

2. Use the "crit1" loop:

```
repeat
    n := n + 1;
    sum := sum + 1/n;
until n = 1000;
writeln('Euler's constant = ', sum - ln(1000));
```

You can check that the resulting number is 0.577 . . . .

3. The infinite sum of inverse fourth powers is sqr(sqr(pi))/90.

4. This is the main method for finding integrals. The only common changes from this algorithm are better geometric approximations to the function, such as replacing rectangles with trapezoids. The integral in the problem has the exact value pi/4.

5. Since the function of x represented by infinitely many summands is a so-called *square-wave*, the numbers obtained with 1000 summands and various x will tend to be either plus or minus pi/4. The peculiar property of the sum is that it hardly changes as the indicated x values are input. The sum flips over to −pi/4 when x crosses the value pi, and so on.

6. The second alternative is

```
n := 0;
sum := 0;
repeat
    n := n + 1;
    sum := (sum + n * n) mod 7;
until n = 52;
writeln(sum);
```

What is so much better about the second option, that of doing mod 7 on each separate summation, is that the sum of the squares will be large. If you were to investigate the mod 7 value of the sum of the first 1,000,000 squares, you might have to use the second option, which keeps all values of "sum" in the set 0, 1, ..., 6; so as not to exceed your computer's largest integer constraint.

APPROXIMATING LIMITS

In the program "crit3" in Figure 2.3, the "until" statement can be paraphrased as "until the sum so far computed is within $1.0e - 6$ of its limit". *Limit* in this case means the sum evaluated for infinitely many terms. One thing that often happens in programs is that the loop that evaluates the sum or limit in question "blows up," that is, has some variable or variables exceeding machine limits. The following example illustrates a potential blowup. Consider the limit

$$\lim_{n \to \infty} \frac{n!\, e^n}{n^n \sqrt{n}}$$

Here, $n! = n * (n - 1) * (n - 2) * \cdots * 2 * 1$ and e is the natural logarithm base ($2.718281828\ldots$).

The wrong way to evaluate the limit is as follows:

Wrong: Compute $n!$ and e^n, then compute n^n and sqrt(n), then multiply the first two, then divide by the second two. Keep doing this for larger and larger n.

The only thing right about this approach is the last phrase because you must evaluate for large n eventually. What is wrong is that n^n gets far too large far too quickly, as do $n!$ and e^n. Is there any way around this problem? Yes. An expression such as

$$\frac{1000!\, e^{1000}}{1000^{1000}}$$

is really not very large (it is about 100) so you should be able to find it with a program. Figure 2.4 shows a *right* way to find the limit (e.g., to $n = 100$) which will be accurate to a few places. The key to this correct approach is to realize that the expression to be evaluated is, except for the relatively weakly growing $\sqrt{n}$ term, a product of n reasonably small ratios. For example,

$$\frac{4!\, e^4}{4^4} = \frac{4e}{4} * \frac{3e}{4} * \frac{2e}{4} * \frac{e}{4}$$

In the program "stir" in Figure 2.4 the corresponding statement is

```
g := g * j * e/n;
```

```
program stir(input, output);
const
     n = 100;
     e = 2.718281828;
var
     j: integer;
     g: real;
begin
     g := 1;
     for j := 1 to n do begin
        g := g * j * e / n
     end;
     writeln(g / sqrt(n))

end.
```

Figure 2.4

This statement prevents g from exceeding machine bounds as long as n is less than some number of millions, depending on your particular machine.

For many limits you can simply compute the expression, without regard for control of the magnitude of the numerator or denominator. As with all programming problems, your approach will be context-dependent.

Exercises

1. Print out a table of the values of the function

$$S(x) = \sum_{\substack{n=1 \\ n \text{ odd}}}^{21} \frac{(-1)^{(n-1)/2}x^n}{n!}$$

for x going from 0 to 20 in steps of 0.5 but formatted side by side with values of sin(x), as follows:

x   S(x)   sin(x)

Remark on whether S(x) is a good approximation to the sine function.

2. Use the product

$$\Pi = 2 \frac{2 * 2 * 4 * 4 * 6 * 6}{1 * 3 * 3 * 5 * 5 * 7} \cdots$$

to obtain a value for pi. Use a criterion similar to that of "crit2" in the text; namely, compute the product until the number $n * n/[(n - 1) * (n + 1)]$ is within $(1e - 7)$ of unity.

3. Find by computer the limit of $[1 - \cos(x)] / \mathrm{sqr}(x)$ as x approaches zero.

4. Write a program to estimate the nested radical

$$\sqrt{2 + \sqrt{2 + \sqrt{2 + \cdots}}}$$

5. Find a point x such that the curve of $x^x = \exp[x * \ln(x)]$ does not essentially change when x changes by some small number dx. This is called a *critical point* of the curve; it is a point with zero derivative. What is the limit of x^x as x approaches zero?

*6. Find the limit of the power ladder

$$\sqrt{x}^{\sqrt{x}^{\sqrt{x}^{\cdots}}}$$

*7. Design a program to find the limit

$$\lim_{n \to \infty} \exp(-n) \sum_{k=0}^{n} \frac{n^k}{k!}$$

Careful! Some will tell you this limit is unity. This is not so, as your computer can suggest.

Answers

1. A good way to compute the sum $S(x)$ is with a loop:

```
sum:= 0;
term:= x;
for m:= 0 to 10 do begin
      sum:= sum + term;
      term:= -term * sqr(x)/((2*m+2) * (2*m+3));
end;
```

Note that when the loop is finished, $S(x)$ is the variable "sum", we have done exactly 11 terms as required, and the minus sign takes care of the required sign-flip for successive summands. The $S(x)$ is an approximation to the infinite series for $\sin(x)$, so that the columns $\sin(x)$ and $S(x)$ you print out will be very similar.

2. The following loop will compute a reasonable value:

```
product:= 2;
k:= 2;
repeat
      next:= (k * k)/((k * k)-1);
      product := product *next;
      k := k + 2;
until abs(next - 1) < le - 7:
writeln(product);
```

Note that the loop termination criterion gives a final precision of about three places, showing that the relation between final-term error and overall expression error is not obvious.

3. The limit is 1/2.
4. The exact result is 2.
5. The exact point for zero derivative is $x = \exp(-1) = 0.37$, as can be obtained with calculus. The limit as x approaches zero is unity.

FUNCTIONS

You are familiar with Pascal's built-in functions such as "sqrt" and "ln". But suppose you need a result such as a^b or arccos(z). The expression to use for powers is

$$a^b = \exp[b * \ln(a)]$$

If powers appear frequently in a program, however, this expression becomes cumbersome to edit. To avoid difficulties, you might define your own function:

pow(a,b)

that returns the value a^b, computed from exp and ln as indicated. Figure 2.5 shows you how to define such a function.

The function definition begins with a *function header* that declares the function *identifier*, its *type,* and all the *arguments* the function requires and their types. The function header is followed by a "begin-end" block, just as in a program. It is important to realize that you can put in additional variables, constants, and so on, *above* the "begin" but below the header. These are called *local identifiers* and are valid only for the function's block, not for another function or for the main block of the program. We shall return to the idea of local variables later.

It is important also to realize that (a,b) in the function "pow" are *dummy variables* in the sense that they have nothing to do with variable declarations

```
program raise(input, output);
var
    x, y: real;

    function pow(a, b: real): real;
    begin
        pow := exp(b * ln(a))
    end; (* pow *)

begin
    read(x);
    read(y);
    writeln(pow(x, y))
end.
```

Figure 2.5

elsewhere in your program. A program containing in its main block

```
x:=2; y:=3
writeln(pow(x,y));
```

will print out the value 8.0. But it is not so obvious what will happen if you have

```
b:=2; a:=3
writeln(pow(b,a));
```

Will the function get confused because you have transposed a, b in the "writeln" statement? Not at all. The result will still be 8.0 on your printer. The placement of (a,b) in the function definition determines *where* to pick up variables when the function value is computed. The function block does not care at all what you call the variables outside in the main program or outside in other functions. Clarify in your mind what dummy or local variables do. Experiment with your own programs.

The situation is more complicated when a global variable is used within the block (along with, for example, a and b, so there would be three variables used in the function block). A *global variable* is one that is defined throughout a program by being declared in the main "var" sequence at the top of the program. The dummy, or local, variables are signified by a "var" declaration within a procedure or function block. You may wish to consult Chapter 1 references for complete descriptions of the roles the variables play.

Exercises

1. Improve the program "raise" in Figure 2.5 by doing a *forever loop* that allows you to type in x, y continually. One way to do this is to put a "repeat" above the "read(x)" statement and to put "until $0 = 1$" below the "writeln" statement. (In rigorous Pascal implementations you can go "until false".) Improve the program further by handling nonpositive x values as follows. Modify the function block (not the main block) by forcing $0^0 = 1$ (see exercise 5, page 24) and any illegal values of x to give an error message of your own devising. Test your methods by running the program and making sure your forever loop does not stop no matter what x, y you type into the computer.

2. Write function blocks to define functions sinh, cosh, tanh, and sech, where

$$\sinh(x) = (\exp(x) - \exp(-x))/2$$
$$\cosh(x) = (\exp(x) + \exp(-x))/2$$
$$\tanh(x) = \sinh(x)/\cosh(x)$$
$$\operatorname{sech}(x) = 1/\cosh(x)$$

Remember that the function block for tanh(x) should appear *below* those for sinh and cosh because Pascal will only look upward to find functions you call.

Write out a table of sinh(x) and tanh(x) for values of x from 0 to 10 in steps of 0.5.

3. Write a program that has a function block giving the area of a triangle with sides a, b, c. Use Heron's formula

$$\text{area} = \text{sqrt}(s * (s - a) * (s - b) * (s - c))$$

where s is one-half the perimeter. For practice in local declarations, you may wish to have a local variable called "perimeter" whose declaration follows the function identifier line. Use your program to give the area of a triangle with sides 1, 2, 2. What is the area of a triangle with sides 1, 2, 3?

Answers

1. An *error-protected function* is as follows:

```
function pow(a,b:real):real;
begin
     if a > 0 then pow:= exp(b*ln(a)) else begin
          if a < 0 then writeln('illegal arg') else pow:= 0;
     end;
end;
```

This will not, however, correctly handle 0^0.

2. These functions can be written as follows:

```
function sinh(x : real) : real;
begin
    sinh := (exp(x) − exp(−x)) / 2
end;
function cosh(x : real) : real;
begin
    cosh := (exp(x) + exp(−x)) / 2
end;
function tanh(x : real) : real;
begin
    tanh := sinh(x)/cosh(x)
end;
function sech(x : real) : real;
begin
    sech := 1/cosh(x)
end;
```

3. The area of the 1, 2, 2 triangle is 0.968246 . . . .
    The area of the 1, 2, 3 triangle is obviously 0.

3 | Equation Solving

All results of the profoundest mathematical investigation must ultimately be expressible in the simple form of properties of the integers.

Leopold Kronecker
From E. T. Bell, Men of Mathematics (1965)

COMPUTER CALCULUS?

Kronecker, the great skeptic, may have gained in the classic nineteenth century debate with his spiritual enemy Weierstrass on whether the irrational number $\sqrt{2}$ exists if a computer had been asked to resolve the conflict (Bell, 1965). In a digital machine, all numbers are in the form of integers, usually binary (base 2), and a number such as $\sqrt{2}$ can only be an approximation.

One of the primary applications of computers in science is to provide *approximate* solutions to equations that involve the *continuum* of real numbers. In this sense, *derivatives* of true calculus origin are approximated by difference equations, and *integrals* are approximated by finite sums. You have seen examples of these approximations in Chapter 2.

Some important types of equations normally solved in the approximate sense are as follows:

1. *Transcendental equations,* such as $x^4 + \ln(x) = 0$, require that the variable such as x be defined implicitly. You search for a number that "almost" satisfies the equation since the true solution is impossible to express algebraically.
2. *Differential equations,* such as $df/dx = -f^2$, require tables, graphs, or special values as the solution because an entire function, such as f, is being sought.
3. *Polynomial equations,* such as $x^7 - x^6 + 2*x^4 + 3 = 0$, have solutions that can sometimes be expressed as roots.
4. *Linear equations,* such as n equations in n unknowns, require that n numerical solutions be found if they exist.

In this chapter we will concentrate on doing calculus by computer by studying techniques that address themselves to equation types 1, 2, and 3. The linear equations of type 4 are discussed in Chapter 6, and more complex forms of differential equations are discussed in Chapter 8.

Exercise

If you are not familiar with differential calculus, study the Chapter 2 references or take a calculus course. You should be familiar with the concepts derivative, slope, maxima, and minima.

DERIVATIVES

The aim in forming a *derivative* is to compute the *slope*. For example, the equation for the slope of the tangent line at $[z, f(z)]$ is

$$f'(z) = \frac{df(z)}{dz} = \frac{f(z + dz) - f(z)}{dz}$$

where dz is very small (taken to zero for the true derivative). The *second derivative* is the slope of the slope, or

$$f''(z) = \frac{d^2 f(z)}{dz^2} = \frac{df'}{dz} = \frac{f'(z + dz) - f'(z)}{dz}$$

If you insert the definition of the first derivative f' into the numerator of this last expression you get

$$f''(z) = \frac{f(z + 2*dz) - 2*f(z + dz) + f(z)}{dz^2}$$

The numerator is called a *second difference*. To get a computer approximation to f'', you need to compute the three numbers $f(z)$, $f(z + dz)$, and $f(z + 2*dz)$, where dz is small.

The importance of derivatives in scientific problems is that they provide certain information, as follows, concerning the function under discussion:

1. If a function f has a relative maximum or minimum at the point z, then $f'(z) = 0$. The computer would be expected to compute a very small value (close to zero) for f' at the point z.
2. At a *critical point,* that is, where $f'(z) = 0$, the function either has a relative minimum or maximum or the point is an *inflection point* (sometimes called a *saddle point*).
3. It is the second derivative f'' that determines which of the three possibilities in item 2 holds: f'' positive means minimum, f'' negative means maximum, $f'' = 0$ means inflection.

You can acquire skill at using these criteria by studying the program "diff" in Figure 3.1 and doing the following exercises.

```
program diff(input, output);
(* compute first and second derivatives of the function f,
   inspect results for min/max behavior *)
const dz = 0.000001;
var   z: real;

    function f(z: real): real;
    begin
     f := z * z * z * z * z - z
    end; { f }

    function df(z: real): real;
    begin
     df := (f(z + dz) - f(z)) / dz
    end; { df }

    function d2f(z: real): real;
    begin
     d2f := (df(z + dz) - df(z)) / dz
    end; { d2f }

begin
    repeat
     write('z: ');
     readln(z);
     writeln('df = ', df(z));
     writeln('d2f = ', d2f(z))
    until false
end.
```

Figure 3.1

Exercises

1. The function defined in program "diff" in Figure 3.1 is $f(z) = z^5 - z$. Using standard calculus, show that there should be critical points at $a = \pm$ sqrt(sqrt(1/5)). Find the values of the second derivative f'' at each of these critical points.

2. Run "diff" program in Figure 3.1. By trial and error find the best numerical values of the critical points to three decimal places. Record the printed values for f'' for these points.

3. Compare the experimental results in exercise 2 with the theory in exercise 1. Sketch by hand what the function should look like on the basis of the data.

4. Write a program that finds the zeros. That is, find those z in the polynomial

$$f(z) = z^4 - z^3 - 7z^2 + z + 6$$

for which $f(z) = 0$. Do this automatically as follows. In the program, initially set $z = 4$ and then decrement z by the small amount dz in a loop. Your approximate zero happens when $f(z)$ changes sign. Stop when the program has found four zeros.

5. Modify the program for exercise 4 so that it finds all the critical points. Do

this by seeking the zeros of f'. (You will see later in this chapter that there is a much faster way to find zeros.)

6. On the basis of your data from exercises 4 and 5 sketch the function. In Chapter 5 you will learn how to plot the function directly using graphics techniques.

Answers

1. Since $f' = 5*sqr(sqr(z)) - 1$, there are critical points at $z = \pm sqrt(sqrt(1/5))$. Since $f'' = 20 * z * z * z$, the second derivative at the two critical points is $\pm 20 * (3/4$ power of $1/5)$, approximately ± 6.
2. Remember to find two points with zero slope.
3. The graph looks like a sideways S with the right-hand (x positive) portion starting downward from the origin and then curling upward as the graph crosses the positive critical point.
4. Your inner loop should look similar to

```
n := 0;
z := 4;
repeat
    z := z - dz;
    if (f(z) * lastf) < 0 then  (* true when of different signs *)
        begin
            writeln('zero', i, 'at', f(z));
            n := n + 1           (* count through zeros *)
        end;
        lastf := f(z);
until n = 4;
```

The theoretical zeros are located at $-2, -1, 1, 3$ for this polynomial.

5. Only two changes are required: Change the call of f(z) to a call of f'(z), and note that there are at most three zeros of the derivative.

NEWTON'S METHOD

Suppose that a body moves such that its position at time t is given by

$$x(t) = t \; sin(t)$$

This motion has two interesting features. First, the motion corresponds to an oscillator that is driven so as to be *undamped*, that is, it oscillates more wildly as time goes on. Second, the equation cannot be solved in the sense that for a given x, you cannot find a t that gives that x (with a few exceptions). This is a transcendental equation.

What are the special solutions for $x(t) = 0$? To find the solutions, either t or sin(t) must vanish, which is accomplished by solving

$$t = n * 2\pi$$

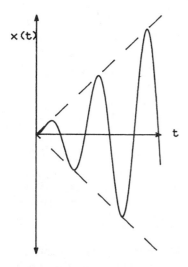

Figure 3.2

where n is an integer. We know when the trajectory $x(t)$ crosses the x axis. The motion is oscillatory as shown in Figure 3.2. What is the velocity for time t? It is given by the first derivative

$$\frac{dx(t)}{dt} = \sin(t) + t \cos(t)$$

Although the velocity is zero infinitely many times, it is not easy to find those times because the equation $0 = \sin(t) + t \cos(t)$ is of the transcendental class.

Thus far you have sketched an approximate trajectory and written down the velocity for any time. To understand the motion more clearly, consider the following example. Suppose that in a physical model for which $x(t) = t \sin(t)$, you want to know the time at which

$$t \sin(t) = 1$$

It is worthwhile to look at the expanded trajectory in the region $0 < t < \pi/2$, as sketched in Figure 3.3. You need to find the special value t' for which $t' \sin(t') = 1$. Use *Newton's method* to get a numerical answer. The method is actually used in handheld calculators to extract square roots and other algebraic functions.

Instead of randomly guessing possible solutions t and seeing which is best, by Newton's method you use a certain measure of error for any one guess in order to predict a next (better) guess. Notice that in Figure 3.3, $t = 1$ is a good initial guess, but that $1 * \sin(1)$—the height of the dashed line—is too small. Newton's method will converge on the value denoted ? in Figure 3.3 as follows. Call the first guess $t0 = 1$, and "shoot a ray" with slope equal to dx/dt evaluated at t0. The process is illustrated in Figure 3.4, where it is shown how to solve $k = t \sin(t)$ for arbitrary k.

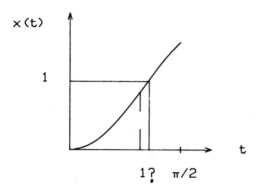

x(t)

1

t

1 ? π/2

Figure 3.3

The constant k in Figure 3.4 is equal to 1 for the present problem, and t_{n+1} represents the next guess obtained with the ray that emanates from the present guess point $(t_n, x(t_n))$. From Figure 3.4 you can see that the numbers involved satisfy

$$\frac{dx}{dt}(t_n) = \frac{k - x(t_n)}{t_{n+1} - t_n}$$

Rearranging this last equation gives the recurrence relation

$$t_{n+1} = t_n + \frac{k - x(t_n)}{(dx/dt)(t_n)}$$

that generates next guesses in an unambiguous way. The program "oscillator" performs in Figure 3.5 Newton's method by incorporating the recurrence relation

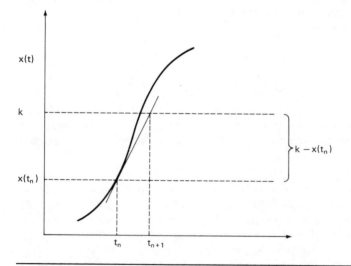

Figure 3.4

```
program oscillator(output);
const
    precision = 1.0e-7;
    k = 1;
var
    last, next: real;
begin
    last := 1;
    repeat
        next := last + (k - last*sin(last))/(sin(last) + last*cos(last));
        last := next
    until abs(next * sin(next) - k) < precision;
    writeln(next)
end.
```

Figure 3.5

for our present case $x(t) = t * \sin(t)$. Notice the all-important initial guess, which has the form

last := 1

The reason this guess is so important is that there are obviously many solutions to $t * \sin(t) = 1$, and Newton's method will seek out a solution that is close to your initial guess.

Newton's method is so rapidly convergent that, for example, when you run the program "oscillator" you almost immediately get the output

1.1147157...

which is a good estimate since this t satisfies $t * \sin(t) = 1.0008$.

Exercises

1. For the trajectory $x(t) = t \sin(t)$ find the *second* time at which $x(t) = 1$. Note that you will have to make a different initial guess t_0 within the program so as not to arrive at the result in the text example.
2. Modify the program to print the value of $t \sin(t)$ for your estimate of t from exercise 1 in order to see if this number is actually close to 1.
3. Use Newton's method to find two different solutions to

    $$t^3 + 2t^2 - t - 1 = 0$$

4. Find the fifth root of 2 by setting $x(t) = t^5$ and finding a number t such that $x(t) = 2$.
5. A potential energy function might take a form such as

    $$V(x) = x \ln(x) \exp(-x)$$

    for positive x. The force on a particle moving in this potential is assumed to be the derivative $-V'(x)$. Find an x such that the force at x vanishes. This problem might arise in a chemistry setting where a complicated interatomic potential must be analyzed.

6. For the function $N(t) = t \exp(-t^2)$, find a solution to $N(t) = 0.01$. The function $\exp(-t^2)$ is ubiquitous in probability theory and appears in countless numerical applications.

Answers

1. An initial guess of 2 will lead you in the right direction.
2. Your program will give you the answer.
3. The three solutions to the equation are 0.802, −0.555, and −2.247.
4. The fifth root of 2 is slightly less than 1.149.
5. This could be accomplished by the following:

```
repeat
    nextx := x - f(x)/df(x);
    writeln(nextx);
    x := nextx;
until abs(f(x)) < 1.0e5;
```

where f is defined as V′ and df is defined as V″. A good initial x is 1.
6. The solution is $t = 2.8185 \ldots$.

DIFFERENTIAL EQUATIONS

Differential equations involve relationships among variables, functions, and derivatives that are to be solved for a function. An example is the following equation that expresses velocity of a mass dropping in a gravitational field:

$$\frac{dx}{dt} = -gt$$

This equation has a simple solution, namely, $x(t) = (\text{constant}) - gt^2/2$. But very often the relevant differential equation is impossible to solve in *closed form*. For this simple velocity equation, you might proceed by declaring variables "new" for position at a time $t + dt$ and "old" for position at time t. Then the differential equation is the difference equation

$$\frac{\text{new} - \text{old}}{dt} = -gt$$

This gives a recurrence relation that can be put into a program such as "fall" in Figure 3.6. The statements

```
t: = 0
old: = height
dt = 0.001
```

translate as

```
program fall(input, output);
const dt = 0.001;
        g = 9.8;         (* acceleration of gravity in meters per second *)
var old, new, t, height: real;
begin
    write('height: ');
    read(height);
    t := 0;
    old := height;
    repeat
      new := old - g * t * dt;
      t := t + dt;
      old := new
    until t >= 10;
    writeln('x(10) =', new)
end.
```

Figure 3.6

Initialize time to zero.
Set first value of x to be "height".
Fix the time increment to be 0.001 second.

Note that initialization of variables is generally important. Also note that dt could have been declared as variable. One reason to do so is that later you may wish to "readln(dt)" to see how precise your difference approximation is to the true dynamical motion.

A more complicated example is the simple-harmonic oscillator equation

$$m \frac{d^2x}{dt^2} = -kx \qquad m, k \text{ constant}$$

This is easily solved in closed form. But almost anything on the right-hand side of an equation except some simple functions may result in a difficult situation. However, to change the right-hand side for the computer, it is only necessary to change one line in your Pascal program; in the present case this would be the line that sets the acceleration variable within the appropriate loop.

There is a clean, universal way to handle second-order differential equations. Think of the second derivative as *acceleration* and the first derivative as *velocity*. Then use the sequence

(determine accel);
vel = vel + accel * dt;
x := x + vel * dt;
t := t + dt;

This is a programming equivalent to Newton's concepts for calculus as it pertains to motion. You should also be aware that more sophisticated and numerically accurate methods of computing derivatives exist (Birkhoff and Rota, 1978). Velocity changes at a rate equal to acceleration, and position changes at a rate equal to velocity. The first line comes from the differential equation giving the second derivative, that is, the acceleration, in some form.

```
program sho(input, output);
const
     dt = 0.01;
var
     k, m, x, t, vel, accel: real;
begin
     write('spring constant: ');
     read(k);
     write('mass: ');
     read(m);
     write('initial position: ');
     read(x);
     write('initial velocity: ');
     read(vel);
     t := 0;
     while t < 1 do begin
          accel := -(k * x / m);
          vel := vel + accel * dt;
          x := x + vel * dt;
          t := t + dt;
          writeln(t, x)
     end
end.
```

Figure 3.7

It is interesting to note that often just one line of your Pascal program contains the law of motion; the rest of the program is *overhead*, that is, it takes care of variables. To see just how much overhead might be required, look at program "sho" in Figure 3.7. It is intended to find the harmonic oscillator motion. Initial conditions are abundant because for *second*-order equations you generally need *two* initial values; in this case "position" and "velocity".

Exercises

1. Run program "sho" given in Figure 3.7 with your own choices for the constants "k", "m", and "initial position". Always set initial velocity equal to zero. From printed output estimate the period of the motion, which is the time required for x to return to its initial value. The theoretical number for period P is

$$P = 2\pi \; \mathrm{sqrt} \left(\frac{m}{k} \right)$$

Verify this relation with various values of "k" and "m". Most importantly, verify the theoretical prediction that P is completely independent of initial x. You may have to change loop conditions to get the proper printout.

2. A radioactive sample of strength N will decay in time according to the differential equation

$$\frac{dN}{dt} = -cN \qquad c \; \text{constant}$$

Solve this equation, for your choices of "c" and of initial number N(0) by iterating the appropriate difference equation. Compare your results to the exact solution $N(t) = N(0) \exp(-ct)$. What is the *half-life* of the sample, that is, the time it takes for N to attain the value N/2? (Give theoretical and numerical answers.)

3. The *pendulum equation* often found in elementary texts is

$$\frac{d^2 x}{dt^2} = -\frac{g}{L} x$$

where $g = 9.8$ meters per second squared and L is the length of the pendulum. Here, x denotes the angle of the bob with respect to the vertical. The equation is equivalent to the "sho" equation in the text but is correct only for small angles x. For larger angles there is error; the true equation is

$$\frac{d^2 x}{dt^2} = -\frac{g}{L} \sin(x)$$

But now the period P is *dependent* on the initial position x(0). According to theory, for $x(0) = \pi/2$, $dx/dt(0) = 0$ the period is $P = 2\pi$ sqrt(L/g) times an absolute constant. Find this constant numerically. It may interest you to know that the constant can be computed from the theory of elliptic integrals. Such numbers are discussed in Chapter 6.

Answers

2. To find the half-life, you want

$$N * \exp(-ct) = (1/2) * N$$

implying that

$$\exp(-ct) = 1/2$$

Therefore, $t = -\ln(1/2)/c$. Your numerical solutions should closely approximate this.

3. The constant is approximately 1.18. It is interesting to note that the numerical solution is rather easy to obtain, while the theoretical solution involves the profound theory of elliptic integrals.

VECTORS

For differential equations involving vectors, such as in descriptions of motion in more than one dimension, you generally relate *components* of the vectors within your programs. Instead of setting up a one-dimensional number x with

```
var x: real;
```

you can declare a variable

var r: array[1..3] of real;

and think of r[1] as x, r[2] as y, and r[3] as z coordinates. The numbers 1, 2, 3 are called the *subscripts* or *indices*. In most implementations of Pascal, you cannot put this sort of declaration in a function, however, and must resort to creating your own type as follows:

type vector = array[1..3] of real;
var r: vector;

Now r can be passed to a function. Some elementary functions are as follows:

1. *Dot product or scalar product*: $r \cdot s = r[1]*s[1] + r[2]*s[2] + r[3]*s[3]$
   (s has also been declared a vector)
2. *Magnitude*: $|r| = \text{sqrt}(r \cdot r)$ (this is the length of a vector r)
3. *Cross product*: $r \times s$ is another vector with components

   cross[1] = r[2]*s[3] − r[3]*s[2]
   cross[2] = r[3]*s[1] − r[1]*s[3]
   cross[3] = r[1]*s[2] − r[2]*s[1]

The angle between two vectors r and s fits into vector analysis problems conveniently in the relationships

$r \cdot s = |r| \ |s| \ \cos(\theta)$

$|r \times s| = |r| \ |s| \ \sin(\theta)$

where theta denotes the angle. Some programmers like to declare types even further, for example, Pascal allows the sequence

type direction = (x,y,z);
        vector = array[direction] of real;
var r, s: vector;

Here the components of r are r[x], r[y], r[z]. Note that x, y, z are not numbers, but symbols, and the declaration of "direction" is an example of *scalar declaration*.
    It is possible to write interesting and instructive programs that make efficient use of scalars. Imagine a program to model the color of a chameleon's skin under various environmental conditions. You might write

type spectrum = (red, orange, yellow, green, blue, violet);
        color = array[spectrum] of real;
var foot, head: color;

Then the color of the chameleon's head might be represented, for example, by a

result that

writeln(head[yellow], head[green]);

gives relative chromatic strengths

1.1   2.13

while printout will show all other color components to be negligible. This would imply a shade of yellow-green as the correct color.

Such use of types is convenient when a program is to have an instructional value or has to be easy to *debug*. In most cases, however, the extent to which you declare arrays is a matter of personal preference.

The programs "printdot" and "printcross" in Figures 3.8 and 3.9, respectively, calculate dot and cross products. There are two features of interest in the cross product program in Figure 3.9. First, since you cannot let a function have an array value, you force the variable c to change within a procedure block. The reliable way to do this is to invoke the second feature, namely, the use of the "var" declaration within the arguments to the function. This assures that c will be changed; recall that functions do not necessarily force change in variables defined outside their own block. For details on the proper declaration of variables as arguments consult Chapter 1 references.

An example of a vector problem is that of a unit mass in the gravitational field of the earth that is allowed to go so high that you must correct for the diminished

```
program printdot(input, output);
(* print the scalar product of two 3-vectors *)
type
    vector = array [1..3] of real;
var
    x, y: vector;

    function dot(a, b: vector): real;
    var
        count: integer;
        sum: real;
    begin
        sum := 0;
        for count := 1 to 3 do
            sum := sum + a[count] * b[count];
        dot := sum
    end; { dot }

begin
    write('enter the three components of the first vector: ');
    read(x[1], x[2], x[3]);
    write('enter the three components of the second vector: ');
    read(y[1], y[2], y[3]);
    writeln('the dot product is:', dot(x, y))
end.
```

Figure 3.8

```
program printcross(input, output);
(* print the outer product of two three-vectors *)
type
    direction = (x, y, z);
    vector = array [direction] of real;
var
    a, b, c: vector;
    count: direction;

    procedure cross(p, q: vector; var r: vector);
    begin
        r[x] := p[y] * q[z] - p[z] * q[y];
        r[y] := p[z] * q[x] - p[x] * q[z];
        r[z] := p[x] * q[y] - p[y] * q[x]
    end; { cross }

begin
    write('enter the three components of the first vector:');
    read(a[x], a[y], a[z]);
    write('enter the three components of the second vector:');
    read(b[x], b[y], b[z]);
    cross(a, b, c);
    for count := x to z do
        write(c[count]);
    writeln
end.
```

Figure 3.9

field far away from the earth's surface. The law of motion is

$$\frac{d^2 r}{dt^2} = \frac{-GM}{|r|^3} r$$

where r is a vector, G is the universal constant of gravitation, and M is the mass of the earth. Use the scheme in the section titled *Differential Equations* to solve the differential equation, but compact the algorithm by using the following procedure that updates various vectors:

```
procedure update(var c: vector; d: vector);
    var i: integer
    begin
        for i := 1 to 3 do c[i] := c[i] + d[i]*dt;
    end;
```

It should be clear that this procedure allows you to update velocity and position if you write

```
update(vel,accel);
update(r, vel);
```

and "r, vel" are assumed to have been declared as vector types. The two calls to "update" are equivalent to the sequence

```
vel := vel + accel*dt;
r := r + vel*dt;
```

Such statements involving vectors cannot be done in Pascal, which is why the "update" procedure has been explained.

It is interesting that the ideas just discussed are enough to create a program that models the orbiting of a mass around a center of attraction. The motion of a planet in our solar system can be so modeled. Of course, perturbations from the other planets must be handled with additional terms in the law of motion. For example, the program "orbit" in Figure 3.10 asks for input of initial components of position and velocity and then models the motion of the body until certain conditions are *not* met (see the following exercises).

```pascal
program orbit(input, output);
const
    dt = 100; (* time increment in seconds *)
    maxt = 1.0e7; (* time limit *)
    g = 6.672e-11; (* gravitational constant *)
    m = 5.98e24; (* mass of earth *)
    re = 6371030; (* radius of earth *)
type
    vector = array [1..3] of real;
var
    accel, vel, r: vector;
    t: real;
    i: integer;

    function mag(a: vector): real;
    var
      i: integer;
      sum: real;
    begin
      sum := 0;
      for i := 1 to 3 do
          sum := sum + sqr(a[i]);
      mag := sqrt(sum)
    end; { mag }

    procedure update(var c: vector; d: vector);
    var
      i: integer;
    begin
      for i := 1 to 3 do
          c[i] := c[i] + d[i] * dt
    end; { update }

begin
    write('initial r components: ');
    readln(r[1], r[2], r[3]);
    write('initial vel components: ');
    readln(vel[1], vel[2], vel[3]);
    t := 0;
    while (mag(r) > re) and (t < maxt) do begin
      for i := 1 to 3 do
          accel[i] := -(g * m * r[i] / (mag(r) * sqr(mag(r))));
      update(vel, accel);
      update(r, vel);
      t := t + dt;
      writeln(mag(r), t)
    end
end.
```

Figure 3.10

Exercises

1. Write a program to build a molecule of carbon tetrachloride (CCl_4). Note that the molecule is tetrahedral with four chlorine (Cl) atoms as vertices and one carbon (C) atom in the center. Work with three-dimensional vectors, where every vector is the location of an atom. Start by placing three chlorines in an equilateral triangle at (x, y, z) coordinates as follows:

$(1,0,0)$

$(-1/2, \sqrt{3}/2, 0)$

$(-1/2, -\sqrt{3}/2, 0)$

Write the program so that it will print out the distance between some pair of these atoms. Then modify the program so that it places a fourth Cl atom somewhere on the z axis such that the distance from the fourth atom to each of the original three atoms is the same as the distance between any pair of the original three. That is, you want to make a regular tetrahedron. One way to do this is to choose positions of the form (0,0,z) for the fourth chlorine then to change z by small increments until all pairwise distances for the four Cl atoms agree.

2. Find the coordinate z' such that a C atom at (0,0,z') is equidistant from the four Cl atoms.

3. Use the program to compute the bond angle, that is, the angle between any two rays leading out from the C atom. In such problems it is a good idea to use dot product formulas to compute cos(theta). *Question:* How do you get theta if you know its cosine? *Answer:* You work out how arctan, an existing Pascal function, is related to arccos.

4. A crystal of a certain salt is formed from small units that appear as parallelepipeds (leaning rectangular boxes). The parallelepiped can be defined by three vectors a, b, c. The volume is generally given by

$$V = |(a \times b) \cdot c|$$

Work out the volume of a unit that has a = (1,0,0), b = (0,1,0), c = (0.2,0.2, 0.8) where all components are in angstroms.

5. In the program "orbit" in Figure 3.10, what is the geometrical significance of the function "mag"?

6. In the program "orbit" in Figure 3.10, state in words what physical conditions are tested in the "while" statement.

7. In the program "orbit" in Figure 3.10, consider motion along the x axis, starting at initial position (components of r)

6371031   0   0

with initial velocity of the form (x component only)

v   0   0

This input of six numbers (you choose v variously) corresponds to the launching of the projectile from the earth's surface. Find the approximate *escape velocity* from the earth's surface, that is, the velocity that just allows complete escape from the earth.

8. Since the time of Newton and Kepler we have known that bodies orbit in ellipses. Figure out a way to input data for initial "r, vel" to get a printout verifying this fact. You may not wish to print out data on every loop pass. Therefore, you may print out every 20th pass of the updating loop as follows:

```
counter := counter + 1;
if (counter mod 20) = 0 then writeln (etc . . .);
```

where "counter" has been declared as integer. Later you will learn how to write graphics programs to plot such orbits.

9. Verify the law of *conservation of energy* as follows. In each pass of the "orbit" loop in exercise 8, print out the total energy

$$E = \frac{1}{2} \, |\, vel\,|^2 - \frac{g*m}{|\, r\,|}$$

and see that it remains reasonably constant for typical orbits.

Answers

1. The distance between the Cl atoms is sqrt(3). To find where to put the fourth Cl atom, it is helpful to have a function that returns the distance between two points, i.e.,

```
function distance(a,b: vector): real;
var
    diff: vector;
    i: integer;
begin
    for i := 1 to 3 do
        diff[i] := a[i] − b[i];
    distance := mag(diff);
end;
```

where you have also written out a "mag" function giving the length of a vector. Then you move the fourth Cl atom along the z axis until all pairwise distances equal sqrt(3). The result for the proper z position of the fourth Cl atom is ± sqrt(2), that is, ± 1.4142 . . . .

2. Use the same idea as in exercise 1 except note that there there are now four neighboring atoms. The exact answer is $z' = (1/4)$sqrt(2).

3. Your results should compare favorably with the exact angle of

$2 * \arctan(\mathrm{sqrt}(2)) = 1.9106\ldots$

which is the famous 109 degree bond angle.

4. The volume is 0.8 cubic angstrom.
5. The function "mag" returns the length of a vector. In the program it represents the height of the satellite above the center of the earth.
6. The condition tested in the "while" statement is the satellite above the surface of the earth (i.e., has not crashed) and the experiment still running.
7. You should observe an escape velocity of approximatly 11,000 meters per second.
8. Your program should exhibit the distance r from the central body changing in time, but periodic (eventually returns to its starting value).
9. You will need a statement such as

```
writeln((1/2) * sqr(mag(vel)) - g * m/mag(r));
```

Note that the energy will not be strictly constant but will vary by a small amount.

MATRICES

A *matrix* can be declared as follows:

```
type matrix = array[1..3,1..3] of real;
var a: matrix;
```

After this declaration you can refer to the elements of the matrix with two indices each. For example, if you write a loop such as

```
for i := 1 to 3 do begin
      for j := 1 to 3 do begin
            a[i,j] := i + 2 * j;
      end;
end;
```

you have assigned to the matrix the values

```
3  5  7
4  6  8
5  7  9
```

This is consistent with the definition that the first index is *row* and the second is *column*. Remember that matrices are conventionally defined as having elements

```
a[row, column]
```

so that a[2,3] above is 8 (row 2 column 3), and so on.

One good application of matrices is in solving simultaneous linear equations. The details of this application are discussed in later chapters, but now you will become familiar with the *matrix library* (Appendix B). First, the procedure `readmat(3,3,a)`

will allow you to input the $3 \cdot 3 = 9$ values of matrix a. Similarly,

`writemat(3, 3, a)`

will output to your terminal the three rows of a.

If you want to solve the equations

$$2x_1 + 4x_2 = 64$$
$$13x_1 - 2x_2 = -4$$

for x_1, x_2 you can proceed in various ways. If you set up a *coefficient matrix* containing the coefficients

$$a = \begin{bmatrix} 2 & 4 \\ 13 & -2 \end{bmatrix} \qquad \text{a is a matrix}$$

and a *constants vector* consisting of the right-hand sides of the simultaneous system

$$c = \begin{bmatrix} 64 \\ -4 \end{bmatrix} \qquad \text{c is a vector}$$

you can solve for the pair (x_1,x_2) as the vector x by simply calling the library procedure

`solve(2,a,c,x);`

where the "2" refers to the *dimension* of the problem (number of equations to be solved). The vector x will be changed to the answer when this procedure is called.

Another use of matrices is the modeling of rotations. The matrix

$$\begin{bmatrix} \cos u & -\sin u \\ \sin u & \cos u \end{bmatrix}$$

is a *rotation matrix* that rotates vectors (x_1,x_2), by an angle u, counterclockwise in the conventional x-y plane (x_1 is x, x_2 is y). The operation of *matrix multiplication,* which includes the special case of rotation, is

$$\begin{bmatrix} a_{11} & a_{12} \\ a_{21} & a_{22} \end{bmatrix} \begin{bmatrix} x_1 \\ x_2 \end{bmatrix} = \begin{bmatrix} a_{11}x_1 + a_{12}x_2 \\ a_{21}x_1 + a_{22}x_2 \end{bmatrix}$$

In other words, a matrix times a vector is a vector. There is a procedure for doing this multiplication in the matrix library. You would declare

```
var a: matrix;
    x,w: vector;
```

where "matrix" and "vector" types have been defined as previously, and call the procedure

mvprod(2,2,a,x,w);

which sets w equal to the product of a and x, as defined above.

Exercises

1. Write a program that declares matrix, vector types, and enough variables to solve the simulaneous x_1, x_2 equations on page 47. Use as many library procedures as you can; definitely include

    readmat (to get the coefficient matrix)
    readvec (to get the constants vector)
    writevec (to output the solution)

2. Write a program that allows input of a two-vector (x_1, x_2) and an angle u and prints out the rotated version of (x_1, x_2), the *magnitude* of this new version, and the magnitude of the *original* pair. Explain why the magnitude is essentially independent of the angle u.

Answers

1. When you apply your methods to the two simultaneous equations on page 47, you must obtain x1 = 2, x2 = 15.
2. Part of your program should look as follows:

    ```
    m[1, 1] := cos(u);
    m[1, 2] := −sin(u);
    m[2, 1] := sin(u);
    m[2, 2] := cos(u);
    ```

    where m is your rotation matrix. The magnitude will not change as the angle u is chosen variously because the magnitude is just the length of the vector being rotated. We say that magnitude is *invariant* with respect to rotations.

4 | Probability Models

It therefore seems that Einstein was doubly wrong when he said that God does not play dice. Consideration of particle emission from black holes suggests that God not only plays with dice but that sometimes he throws them where they cannot be seen.

S. Hawking (1977)

RANDOM REAL NUMBERS

Many programs using probability concepts, such as that of rolling dice, require random numbers to be generated. Since computers are deterministic, it is hard to see how they can produce truly random numbers. In fact they cannot. Instead they use *pseudorandom* numbers, which are generated in such long sequences that you can hardly guess the number following a given number. Advanced theory of pseudorandom numbers is discussed in Knuth (1971). A typical statement that will print out a random real number is

```
writeln(random(1));
```

or

```
writeln(random);
```

This is usually a real number between 0 and 1, equidistributed over that interval. Further calls of the "random" function in the program result in more members of the long pseudorandom sequence. However, in most systems you get the same sequence printed out every time you run the program. Because this violates the very notion of random, systems have an option to *seed* their internal random number generator. An example of a seed operation is a *single* statement at the beginning of the program block, such as

```
k := seed(wallclock)
```

Note that such statements vary radically from system to system. In the example given, the *real time* ("wallclock" is in seconds) of day is always changing, so the next time you run your program the printed random numbers will start at a different place. You should find out how to seed your system's generator, especially for programs that require an unpredictable start, and therefore an unpredictable continuation, of random numbers.

Computer-generated random numbers are quite versatile. A coin flip, for example, can be modeled with

```
if random(1) < 0.5 then heads := heads + 1
   else tails := tails + 1;
```

and every time this statement is executed, exactly one of the declared integers—heads or tails—will increment with equal probability. *Question:* What is being modeled if "0.5" is changed to some other number? *Answer:* A weighted coin.

More sophisticated distributions than the basic equidistribution can be obtained for general application in probability models. See Appendix C for special random number generation.

Exercises

1. An example of a simple biological model is as follows. Write a program to model sex of offspring, declaring integers "m" and "f" for number of males and females, respectively. Assume that births of males and females are equally likely. Generate "m", "f" for 100,000 offspring and print out these two integers. Verify that running the program with no seed gives the same result but that running it with seed gives different results every time. (Of course, if you ran it, e.g., several trillion times you would get a repeat because of the pseudorandom character of the generator.)

2. Modify the program in exercise 1 to print out the length of the longest unbroken string of females. Then modify this program to provide an experimental answer to the following question: If New York City is thought of as the home of 2,000,000 families, each having two parents plus three children, how many families have three daughters and no sons?

3. Model *particle diffusion* as follows. A particle starts at the origin (0,0) in two dimensions and always jumps in a direction north, west, south, or east with the equal probability of 1/4. This means that after one jump it is equally probable that the particle is at (0,1), (1,0), (−1,0), or (0,−1). Write a program to jump a large number of times, and print out the distance of the particle from its starting origin, namely $r = sqrt(sqr(x) + sqr(y))$, after all jumps are completed. Theory predicts that the expected value of r is $K * sqrt(n)$ after n jumps, where K is an absolute constant. Modify the program with a seed and start over enough times so that you get a plausible experimental value for K.

Answers

1. You need a statement such as

```
if (random(1) > 0.5) then
   males := males + 1
else
   females := females + 1;
```

which must be passed 100,000 times. Then you need the statement

writeln(males,females)

2. A correct solution to the first part of exercise 2 is to rewrite the loop of exercise 1 to read

```
if random(1) > 0.5 then begin
     males := males + 1;
     fcount := 0;
     end else begin
          females := females + 1;
          fcount := fcount + 1;
     end;
if fcount > fmax then fmax := fcount;
```

After initializing fmax and fcount to 0, you pass this whole segment 100,000 times. fmax is the running maximum pure female string length, and fcount is the number of females since the previous male.

For the second part, observe that the probability of a no son family is 1/8. Then the loop

```
for n := 1 to 2000000 do
     if random(1) < 1/8 then count := count + 1;
```

will establish "count" as the required number of families. You can go a step further and flip the coin three times for each family, checking for all daughters, and so on.

3. The theoretical value for K is sqrt(pi/4).

RANDOM INTEGERS

Though the function "random" returns a real number between 0 and 1, this does not prevent you from using the same function to generate *random integers*. If k is any positive integer, the expression

1 + trunc(k * random(1));

is a random integer between 1 and k inclusive. Each integer $1, \ldots, k$ is equally likely. Look at the expression to convince yourself that every integer $1, \ldots, k$ is possible but not 0 nor $k + 1$.

An example of problems requiring random integer generation is the *shuffling problem*. The idea is to rearrange the elements of an array in a random way. We will illustrate this problem with the example of shuffling a deck of cards. More scientific examples of shuffling are as follows:

1. Shuffling of genes for modeling genetic drift (see Chapter 9)

```
procedure shuffle(var x: deck);
(* call once to shuffle deck x *)
var
    m, n: integer;

    procedure swap(a, b: integer; var y: deck);
    var
        temp: integer;
    begin
        temp := y[a];
        y[a] := y[b];
        y[b] := temp
    end; { swap }

begin
    for m := 52 downto 1 do begin
        n := 1 + trunc(m * random(1));
        swap(m, n, x)
    end
end; { shuffle }
```

Figure 4.1

2. Modeling molecular collisions or other microscopic statistical phenomena involving integer numbers of events (see following exercises)
3. Testing number-theoretical hypotheses involving integers (see following exercises)

For our deck of cards example, we define our type "deck" as

```
type deck = array[1..52] of integer;
var x: deck;
```

The components of x will be integers 1, 2, . . . , 13; each integer will occur four times. There are no suits for the moment. Consider the procedure "shuffle" in Figure 4.1. It is sometimes called an *include file*, since it may be saved on your system and included in programs as you wish. Notice the *algorithm* that appears as the main block of the procedure "shuffle" (at the end of the listing). It calls the procedure "swap". Note that the "vars" of the overall procedure are written before those of the inner procedure.

Exercises

1. Describe what procedure "swap" does.
2. Write a short description of what the actual shuffling algorithm (beginning "for m := 52") does to the deck. What is the probability that the integer value originally in x[52] is now in x[12]?
3. Practice writing moderate-length programs having *purely procedural main blocks*. Start by creating a procedure that begins

    ```
    procedure setup(var x: deck);
    ```

    which is to initialize the x array to [1,..,13, 1..,13, 1..,13, 1..13], and then create a procedure "deal" that will print the first five cards of the deck.

4. Write and run a program that has the main block (at the end of the listing)

```
begin
        setup(x);
        shuffle(x);
        deal(x);
end.
```

Run this and see that you get five cards and that if you seed the random generator you get a different set on future runs.

5. The program in exercise 4 actually does very little. A more realistic task is to modify the main block to solve the following question: How many hands of five do you need before you get four of a kind (such as 10,10,10,10)? The problem of when to shuffle is left to you.

6. Suppose that a glass plate is bombarded by 100,000 silver atoms and they all adhere to the plate. The glass is then broken into 36 pieces, all of equal area. Assuming that the arrival of atoms is uniformly random over the surface, write a program that models the experiment. The goal is to list 36 integers, each representing the number of atoms that struck a different piece of glass. The program should begin

```
program sterngerlach(output);
var glass: array[1..36] of integer;
```

so that glass[j] is to be the number of atoms striking the jth piece. You must loop 100,000 times in your program, each pass of the loop incrementing a random piece by one atom. Later you will be able to analyze the output statistically so you may wish to direct the final list of 36 integers to a disc file. This problem is patterned after the classic *Stern-Gerlach experiment* (Tipler, 1969).

7. If 100 committee members each write their names on a piece of paper and mix them in a hat and each member then draws a name, what is the probability that nobody gets his or her own name? Assume no two members have the same name.

*8. Estimate by random integer generation the probability that two random integers have a common divisor greater than 1 (Hardy and Wright, 1965).

*9. Write a procedure that models the *Bernoulli distribution* as follows. A new function should be defined:

```
function bernoulli(n: integer): integer;
```

which returns a random integer from 0 through n, having value j with probability

$$P_j = 2^{-n} \frac{n!}{j!(n-j)!}$$

These numbers add up to unity as j ranges 0, 1, ..., n. The numbers bernoulli(n) $-$ n/2

will be distributed in a classic bell-shaped histogram.

Answers

1. "swap" exchanges the values in two positions of deck y.
2. The last card is swapped with a card of random position, then the next to last card is swapped with a random card of the first 51 cards, and so on. The probability is 1/13, because the shuffle is random and there are 13 card values.
3.

```
procedure setup(var x: deck);
    var j: integer;
    begin
        for j := 0 to 51 do x[j + 1] := j mod 13 + 1;
    end;
procedure deal(var x: deck);
    var j: integer;
    begin
        for j := 1 to 5 do writeln(x[j]);
    end;
```

4. Incorporate the procedures from the text and exercise 3.
5. There is nothing wrong with shuffling after every hand, although the search for four of a kind will be slow. For the constant-shuffle approach a solution is

```
hands := 0;
repeat
    shuffle(x);
    deal(x);
    hands := hands + 1;
until four;
writeln(hands, 'hands were dealt to get four-of-a-kind');
```

The boolean-valued function "four" returns "true" value if and only if there are four or more identical cards in a hand of five. A little thought reveals that there are four equal cards in a set of five if and only if the total number of unequal pairs (out of ten pairs) does not exceed four. Therefore you can define

```
function four: boolean;
    var i, j, u: integer;
    begin
        u := 0;
        for j := 1 to 4 do begin
        for i := j + 1 to 5 do
            if x[i] < > x[j] then u := u + 1;
        end;
```

```
(* u = number of unequal pairs *)
if u > 4 then four := false else four := true;
end;
```

6. The key loop is

```
h := 1 + trunc(36 * random(1));
glass[h] := glass[h] + 1;
```

which you pass 100,000 times.
7. Theoretically, 1/e is the probability of zero matches. This is about 37%.

MONTE CARLO METHODS

If a function f(t) is defined for $0 < t < 1$ and you wish to compute the integral

$$I = \int_0^1 f(t)\, dt$$

it is useful to observe that I is the average value of f(t) over the interval. It is therefore possible to sum many values of f(t) for random t and divide the sum by the number of samples. Such a technique is embodied in the program "reno" in Figure 4.2, where the function $f(t) = t^3 - t^2 - t + 1$ is integrated.

The program will generate a *Monte Carlo integral* of your chosen function. *Question:* What is the error in this procedure? *Answer:* The error in this kind of integral is "1/sqrt(nsamples)", as can be derived in probability theory. In exercise 3, page 50 you can show that the expected distance of a randomly walking particle from the origin is proportional to the square root of the total number of steps taken. In a similar way, the expression

area − (nsamples) * I

```
program reno(output);
const
    nsamples = 10000;
var
    area: real;
    sampl: integer;

    function f(t: real): real;
    begin
        (* an arbitrary function*)
        f := t * t * t - t * t - t + 1
    end; { f }

begin
    area := 0;
    for sampl := 1 to nsamples do
        area := area + f(random(1));
    writeln(area / nsamples)
end.
```

Figure 4.2

(recall I is the true integral) is a random walk with "(nsamples)" number of steps. The number printed out—"area / nsamples"—does have the following error associated with it:

area / nsamples − I

which decreases, by the random walk analogy, in proportion to "1/sqrt(nsamples)".

Note that for integration limits other than 0, 1 (call them "low" and "high") the term "f(random(1))" must be replaced with

f(low + (high − low) * random(1));

so that the argument to "f" has the proper range. The integral's output value is then

area * (high − low) / nsamples;

In all such cases the Monte Carlo technique converges slowly, but any function you can define in a program can be integrated.

Monte Carlo methods are especially useful for the evaluation of many-dimensional integrals that appear in chemistry and physics. In such cases you can choose random points having n coordinates—that is, you can choose n random real numbers each with its own special range—and sum the particular function over many such points.

Exercises

1. Perform a Monte Carlo integration for

$$\int_0^{\pi/2} \cos^2 t \, dt$$

and compare your result with the exact integral.

2. Perform a Monte Carlo integral for

$$\int_{-3}^{+3} \exp(-t^2) \, dt$$

and compare your result with the theoretical answer obtained when the domain of integration is expanded to $(-\infty, \infty)$.

3. Perform a different Monte Carlo as follows. For the function $f(t) = 1/(1 + t * t)$ notice that the integral

$$\int_0^1 f(t) \, dt$$

is the area under f(t) that also lies in the *unit square* with corners (1,0), (1,1), (0,1), and (0,0). Sketch the area that is within the unit square. Write a program to select random pairs (x,y) within the square, and count how many pairs fall under the curve f(t). Use this number to estimate the integral. Compare your estimate with the exact result (see exercise 4, page 20).

4. By choosing various total numbers of points dropped in exercise 3, verify qualitatively that the process seems to converge on the exact result with an error of order "1/sqrt(total # dropped)".

*5. Write a program to find the volume of an n-dimensional unit ball using the Monte Carlo technique. The volume will be the number of points x with the sum,

$$\sum_{j=1}^{n} x_j^2 < 1$$

divided by the total number of points in the *containing region* (region from which selection of x_j is made). A convenient containing region is the unit n cube defined by

$$-1 < x_j < 1 \qquad j = 1, \ldots, n$$

having volume 2^n.

*6. *Buffon's needle* is a line of length L that is repeatedly dropped onto a floor consisting of ruled parallel lines separated by spaces that are of width L. The probability of crossing a line is known to be 2/pi. After many drops you should have

$$\frac{\text{Number of crossings}}{\text{Number of drops}} = \frac{2}{\text{pi}}$$

The task is to write a program that drops many Buffon needles and finds a value for pi. A special aspect of this problem is to write the program *without recourse to the value of pi* at the beginning. For example, you might be tempted to drop a needle so that its center is at (x,y) where

x := random(1); y := random(1);

and one end is at

(x + L * cos(theta) / 2, y + L * sin(theta)/2))

where theta is a random angle. The difficulty is that the natural expression

theta := 2 * pi * random(1);

should not be allowed. (A Buffon needle Pascal experiment is shown in Figure 5.13.)

Answers

1. The exact integral is pi/4.
2. The integral over the whole real line is sqrt(pi). The integral over (−3,3) is the same to about four places.

3. The area is pi/4. Inside your loop you want

```
x := random(1);
y := random(1);
if y < (1/1 + x * x)) then
   hits := hits + 1;
```

4. You need to drop many thousands of points in order to see the qualitative effect of "1/sqrt(n)" convergence.

STATISTICS

Computer programs are often used to support experimental work in the sciences, for example, statistical analysis of experimental data. For a theoretical treatment of statistics, consult a text such as Eason et al. (1980). Suppose the data are $x_1, \ldots, x_{size}$ where the *sample* is the array x and the integer *index* size is the sample size. A typical declaration of the array, allowing up to 100 elements in the sample is

```
const maxsize = 100;
type sample = array[1..maxsize] of real;
var x: array;
```

To obtain a smaller sample, truncate the array by keeping track of another variable "size", declared as integer. *Question:* Can you just change the dimension of the array, that is, the number "maxsize" in the present example, during program execution? *Answer:* No. This is one of the most prevalent complaints about Pascal. In most cases, however, keeping track of the largest index will allow you to get around this problem of lack of dynamical arrays. You compute on the basis of values $x[1], \ldots, x[size]$; and ignore the remaining values $x[size + 1], \ldots, x[maxsize]$.

Two important quantities to be extracted from a data array are mean and error. The *mean* is the arithmetic average:

$$\text{Mean} = \frac{1}{size} \sum_{n=1}^{size} x[n]$$

Error, sometimes called the *standard deviation,* is derived from the formula

$$(\text{error})^2 = \frac{1}{size - 1} \sum_{n=1}^{size} (x[n] - \text{mean})^2$$

The standard deviation is a measure of how far typical data are spread from the mean. The number $(\text{error})^2$ is sometimes called the *variance.* The error is essentially the width of the *normal distribution* (that is, the *bell curve* or *Gaussian curve*), that best fits the data. These and many other statistical functions are described in Appendix C.

One of the most important uses of arrays in data analysis is the determination of the *best-fit straight line* that describes the data. The technique is known as *linear regression.* For given arrays x and y (with the same size) you find that exact pair of real numbers (m,b) such that the sample defined by

$$z[n] := m * x[n] + b; (* n = 1, \ldots, \text{size} *)$$

minimizes the squared error

$$E^2 = \sum_{n=1}^{\text{size}} (y[n] - z[n])^2$$

The problem of finding the best m and best b can be solved completely. The solution can be obtained by using calculus. Differentiate the expression E^2 first with respect to m and then with respect to b. Both of these first derivatives must be zero if the squared error is to be a minimum. When you set both of the derivatives to zero you find

$$m = \frac{x \cdot y - \text{size} * \text{mean}(x) * \text{mean}(y)}{x \cdot x - \text{size} * \text{mean}(x) * \text{mean}(x)}$$

$$b = \text{mean}(y) - \text{mean}(x) * m$$

These values of m and b give the best slope and intercept, respectively, for a straight line running through the field of data points (x[n], y[n]). In the notation above, the dot product x · y denotes the sum

$$x \cdot y = \sum_{n=1}^{\text{size}} x[n] * y[n]$$

by analogy with the vector dot product (see Chapter 3, section titled *Vectors*). Like the mean and error, the best-fit parameters m and b can be calculated using procedures described in Appendix C.

Another useful operation to perform on data samples is *sorting.* There is an enormous need in commercial applications for software that performs the sort procedure. There are many kinds of sorting algorithms that can be used to order a list; some are essentially fastest-possible algorithms. Here we will illustrate a simple sort, not the fastest one.

Assume that you are given a sample x.

1. If the first element x[1] is more than x[2], then swap(1,2,x), else do nothing. Note that swap is used as in the procedure "shuffle" of Figure 4.1 and the section titled *Random Integers.*
2. Check for x[1] > x[3] and swap or not accordingly. Then check x[1] > x[4], and so on, until you have checked x[1] > x[size].
3. Increment the 1 index that was fixed in steps 1 and 2 so that you check x[2] > x[3], then x[2] > x[4], and so on, always swapping or not.

4. Continue to increment the fixed index until you have checked all possibilities up through x[size − 1] > x[size].

The sample x is now sorted so that the elements are arranged from smallest to largest as the index goes from 1 to size. The relevant Pascal statements that achieve steps 1 through 4 illustrate how elegant programs can be. For example,

```
for j := 1 to size − 1 do begin
    for k := j + 1 to size do
        if x[j] > x[k] then swap(j,k,x);
end;
```

Convince yourself that this will sort any sample x before beginning the exercises.

Exercises

The library package for statistics in Appendix C contains procedures and functions that are relevant to the following tasks.

1. Write a program that inputs a list of real numbers and prints out the same list *sorted in decreasing order,* that is, x[1] becomes the largest number in the list.

2. Write a program that prints out the mean and error of a sample that you input. Test this work and save it for exercise 4.

3. Write a program that prints out the best slope and intercept for data points that you enter in the following format:

```
x[1]  y[1]
x[2]  y[2]
x[3]  y[3]
. . .
```

[In most systems there is an *EOF* (*end of file*) character, such as "control-D" or "control-C", to indicate to the program that the list is finished.] Save this work for Chapter 5.

4. For your programs in exercises 1 and 2, use file input instead of manual terminal input. Use the file created in exercise 6, page 53 that contains 36 real numbers, sort it, and print the mean number of atoms per piece and the error. Explain why the mean could have been obtained immediately by a trivial calculation.

We will postpone using a file input instead of terminal input for the best-fit task in exercise 3 until Chapter 5, in which graphics techniques are covered, so that the best-fit line can be drawn.

5. Use the *Poisson distribution function* poiss(x) described in Appendix C in a modeling problem as follows. First, understand that poiss(x) returns a random integer in such a way that the expected value of this integer is x. Furthermore, the number returned from poiss(x) simulates how many atoms would hit a piece of glass if x were expected. In particular, the probability that the integer j is returned is

$$\frac{x^j \exp(-x)}{j!}$$

Now run a simulation of the glass experiment (exercise 6, page 53) by calling poiss(x) 36 times and using what you know x must be on the basis of the reasoning in exercise 4 above.

6. Compare qualitatively the simulation in exercise 5 above with the original experiment in exercise 6, page 53 by inputting the simulation glass data for the programs in exercises 1 and 2. Are the means similar? Are the errors similar? What is the advantage in using "poiss(x)"?

7. In the program in this exercise you will get an intuitive idea of standard deviation. For our purposes, standard deviation is synonymous with error and therefore equal to the square root of the variance.

Using the library function "gauss(mean, variance)" described in Appendix C, write a program to model maximum daily temperature in Portland, Oregon. Assume each day's temperature is a Gaussian distribution with mean 20° Celsius and variance 9. Have the program compute the temperature for 2 years, that is, 730 values. Print out the mean of the numbers (which is about 20°C), the greatest and least temperatures for the time period, and the number of days for which the temperature was at least 2 standard deviations away from the mean (either direction).

Several things are wrong with this type of Gaussian model. First, the implication in the problem is that days are not correlated with each other, yet we know that weather runs in seasons. *Question:* How would you fix this? Second, a gaussian distribution cannot rigorously model a quantity such as temperature because temperatures are constrained to lie within fairly rigid extremes. What is the range of a true gaussian variable?

Answers

1. Use the Pascal statements on page 60, but change $>$ to $<$.
2. Remember that you have library procedures mean and error.
3. You can either write out the expression for the slope and intercept or use library procedures "bestm" and "bestb".
4. Input directing varies from system to system. The trivial calculation for mean is mean := 100000/36, because that is the number that, on the average, should hit a given region.
5. Use x := 100000/36; the results of using the "poiss" function should be qualitatively similar to the Monte Carlo results in exercise 6, page 53. The important statement here is

```
glass[j] := poiss(100000/36);
```

6. The mean and error should be within about 10% of each other. The advantage of using "poiss" is that it is faster than atom-by-atom microscopic bombardment.

7. The mean must be about 20° Celsius. The largest and smallest temperatures depend on your system's random generator. The number of days having 2 standard deviations will be small, that is, about two or three, again depending on your system. There are many ways to correct for the lack of seasonal correlation. One way is to force a correlation between a given temperature and the most recent temperature, for example,

    temp := 0.5 * gauss(20, 9) + 0.5 * tlast;
    tlast := temp;

    with tlast initialized to 20. In general, all temperatures are still about 20° Celsius, but they are correlated day to day. The reason a gaussian distribution is not rigorous in this problem is that the true range of a Gaussian variable is the whole real line, whereas temperatures have definite limits.

HISTOGRAMS

Histograms are used in all sciences for visualization of data. They are best displayed on graphics terminals, but it is important to realize that you can sometimes get useful plots with a regular printer. (We shall be doing many types of graphics in Chapter 5.)

To set up a histogram, allocate to each of N *bins* the number of data points lying in a prescribed interval associated with each bin. Consider the following data:

1.12 1.67 2.30 3.46 1.17 3.99 4.09 1.08
1.56 2.13 4.60 4.50 4.70 3.89 4.98 1.98

You should be able to write a program that decides first how many of these data are in the interval [1,2), in which the punctuation [ ) means that data d satisfies

$1 \le d < s,$

then how many are in the interval [2,3), and so on. A histogram for the problem is

bin	count
1	6
2	2
3	3
4	5

showing, e.g., that five data lie in [4,5). Notice that N = 4 total bins for the problem and the sum of all the bin *counts* is exactly 16, which is the number of data points.

A printer can be used to output the data as follows:

```
bin
  1     XXXXXX
  2     XX
  3     XXX
  4     XXXXX
```

by printing an "X" for each count in the appropriate bin. This approach is more economical than graphics and often just as effective.

Exercise

Write a program for a histogram for a Gaussian sampling as follows. Create 730 temperatures as in exercise 7, page 61, and print out the number of counts in each bin corresponding to 10°C through 29°C. The total number of bins is N = 20, and the printout should be of the form

```
bin     counts
10      (print number of temperatures in [10,11) here)
11      ... etc.
 .
 .
 .
29      ...
```

where you simply ignore (i.e., do not count) temperatures greater than 30°C or less than 10°C. Once you are convinced the counts are correct, change the program to print out strings "XXXXXX . . . " so that you get a picture of a bell curve. You may have to let one symbol "X" represent more than one count to allow the histogram to fit on the paper.

Answer

A procedure that works is

```
procedure histo(count: integer);
var
    i: integer;
begin
    for i := 1 to count do
        write('X');
    writeln;
end;
```

and you can replace "count" in the "for" statement with, for example, "count div 3" to display one "X" for more than one count.

5 | Modeling with Graphics

I think Isaac Newton is doing most of the driving right now.
Major W. A. Anders
(by radio, first circumlunar voyage)

PREPARATION

To anyone already familiar with graphics techniques, the Major's words have special importance. It is wonderful and instructive to see a dynamical law, or for that matter any appropriate law of science, play out its consequences on a graphics screen. Up to now this book has dealt exclusively with printed output, and now we prepare for a genuine change of focus. One word of caution: Graphics is not necessarily the ideal form of output. Whether to use printout, graphics, or some other alternative such as audio output from programs is a task-dependent question. You should ponder this point before going on to creative graphics, so that you do not sacrifice unnecessary time to the recreational aspects of your graphics terminal.

To prepare for the work of this chapter, you must implement certain procedures on your system. Appendix A has explicit procedures written out for you, but you can certainly implement any equivalent procedures for your system. Appendix A includes:

plot(x,y)   Put a point at coordinates (x,y).
draw(x,y)   Draw a straight line to (x,y).
splot(x,y,z,a,b,c)   Place a point in three dimensions, overall orientation (a,b,c)
                     for the resulting screen projection.
sdraw(x,y,z,a,b,c)   Draw a straight line to (x,y,z), to be viewed also in orientation
                     (a,b,c).

By "orientation (a,b,c)" we mean that 3-space is rotated by the *Euler angles* (a,b,c) as described later in this chapter. These key procedures will enable you to do points and lines in such a way that you can draw two-dimensional pictures, such as graphs of functions, histograms, and trajectories, or three-dimensional pictures, such as graphs of surfaces and figures (as in drafting drawings).

Exercise

Familiarize yourself with Appendix A so that you will be able to adapt your own plotting system for consistency with the procedures in Appendix A. You will want, for example, a procedure "plot(x, y)" to put a dot at x, y, where x and y each range from −1 to +1. The procedure "draw" is defined similarly. The functions "plot", "draw", "splot", and "sdraw" are sufficient to do almost all of the problems in this book. Find out whether your system has color capability, refresh capability (for animation), and other special capabilities.

GRAPHING OF FUNCTIONS

Suppose that we have a function f defined on the interval $(-1,1)$. Note that there is the constraint

$$-1 \leq x \leq 1 \; [\text{abs}(x) > 1 \text{ used for } \textit{alphanumeric labels}]$$

on all of the work in this chapter. It is important to know exactly what the limits are for the graphics screen display. Accordingly, we have set limits on the function f itself by

$$-1 \leq f(x) \leq 1$$

The origin $(0,0)$ of the two-dimensional (x,y) coordinate system thus lies at the center of the screen.

You can graph functions as follows:

1. Lift your pencil and place it down on an initial point $[x,f(x)]$.
2. Draw a straight line to the next point $[x + dx, f(x + dx)]$ where dx is sufficiently small that broken lines will give a good visual approximation to the true curve.
3. Continue to increment x by dx until some limit is reached on x.

The program "graphf" performs these steps as shown in Figure 5.1. Notice the procedures "clear", "graph", and "move". The "clear" procedure clears the graphics screen. The "graph" procedure sets up the terminal for the graphics mode. The "move" procedure is really step 1 above. In other words, in a "move" procedure the pencil does not touch the paper in between making dots. *Question:* Would "plot" work instead of move in the program "graphf"? *Answer:* Yes, if your system works such that "plot" always lifts the pencil before making its next dot.

In the program "graphf" it is necessary to insert function definitions such that $f(x)$ is either in the interval $(-1,1)$ or is not plotted. There are also ways to *squeeze* a picture onto the display. One approach is as follows. Define constants such as

```
const xscale = 3;
      yscale = 4;
```

then always "plot" or "draw" according to

```
plot(x/xscale, f(x)/yscale);
draw(x/xscale, f(x)/yscale);
```

where "xscale" and "yscale" are chosen to squeeze the picture properly into the normal plotting region. These *scaling operations* must be changed if the origin is

```
program graphf(input, output);
const
    first = -1;
    last = 1;
    dx = 0.01;
var
    x: real;

#include "plibh.i"                    (* include graphics library *)
    function f(x: real): real;
    begin
    f :=                             (* put your function here *)
    end; { f }

begin
    clear;
    graph;
    x := first;
    move(x, f(x));
    repeat
    x := x + dx;
    draw(x, f(x))
    until x >= last
end.
```

Figure 5.1

not to be in the center of the screen. The most general form of scaling is to define x and y *offsets* such as

xoff = 0.2;
yoff = −1;

and then to

plot(x/xscale + xoff, f(x)/yscale + yoff);

Question: Where is the origin plotted in this general case? *Answer:* It is plotted at "(xoff,yoff)".

Often you may not wish to draw parts of curves that are *off scale* and, more importantly, parts of curves where internal computation would blow up. The program "graphf2" in Figure 5.2 graphs the function $f(x) = \tan(x)$, which has points at which the curve is *off screen* and points at which the curve cannot be legally computed [$\tan(x)$ is not defined for $x = \pi/2, 3\pi/2, \ldots$]. This program has key statements that are described according to their functions as follows:

Statement	Function
"if cos(x) < > 0 then ..."	Checks to see if tan(x) will blow up in the function block.
"if abs(...) > 1 then ..."	Checks so as not to plot anything if the point is off screen (such points *are* plotted at distorted positions if such checks are not made).
"until x > last"	Stops drawing in any case if x exceeds the given limit.

The picture you get from this program is shown in Figure 5.3. It is not difficult to put in axes and, if your system has the option, alphanumeric labels.

```
program graphf2(input, output);
const
    first = -5; (* initial x-value. *)
    last = 5;   (* final x-value *)
    xscale = 5;
    yscale = 6;
    dx = 0.01;  (* x increment *)
var
    x: real;

#include "plibh.i"                  (* include graphics library *)

    function f(x: real): real;
    begin
        f := sin(x) / cos(x)
    end; { f }

begin
    clear;
    graph;
    x := first;
    move(first/xscale, f(first)/yscale);
    repeat
        if cos(x) <> 0 then
            if abs(f(x) / yscale) > 1 then
                move(x / xscale, f(x) / yscale)
            else
                draw(x / xscale, f(x) / yscale);
        x := x + dx
    until x > last
end.
```

Figure 5.2

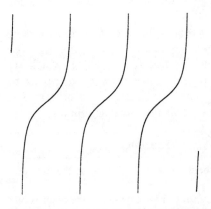

Figure 5.3

Exercises

For these exercises you will need a graphics package equivalent to that in Appendix A, especially the procedures "plot" and "draw".

1. Modify the program "graphf2" in Figure 5.2 so that axes (you do not need to label these) and asymptotes are drawn. The latter should be vertical dashed lines at the positions

$$x = \pm\frac{\pi}{2},\ \pm\frac{3\pi}{2},\ \pm\frac{5\pi}{2},\ \ldots$$

2. Taking into account the problems of blowup and off-screen values, as is done in "graphf2", make a plot of the function $f(x) = 1/x$ over the interval $-10 < x < 10$.

3. Make a plot of a planetary orbit as follows. A polar coordinate representation of the orbit is

$$r = \frac{1}{1 - h\ \cos(theta)}$$

where (r,theta) are related to cartesian coordinates by

$x = r \cos(theta)$
$y = r \sin(theta)$

Let h be defined as less than 1 so the orbit is an ellipse. Plot various ellipses as h is changed. Why is h called the eccentricity?

4. Model the phenomenon of *amplitude modulation* by making a graph of the function

$$V(t) = \cos(2\pi at) + \cos(2\pi bt)$$

where t (time) is graphed on the horizontal axis and V on the vertical axis. It is convenient to have frequencies a = 30 and b = 24 on the first run. By running several cases of (a,b), verify that the frequency of the envelope of the waveform (i.e., the beat frequency) is abs $(b - a)$ when b is nearly equal to a. In Figure 5.4, P is the period of the beat, and 1/P is the beat frequency. You may wish to add axes and tick marks.

    The amplitude modulation can be thought of mathematically as follows. There is a trigonometric identity

$$\cos x + \cos y = 2\ \cos\left(\frac{x+y}{2}\right)\cos\left(\frac{x-y}{2}\right)$$

When this identity is applied to the given V(t), it shows that the wave can be represented by the product of a high-frequency wave (carrier) and a low-

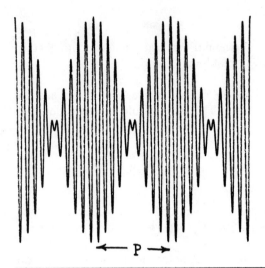

Figure 5.4

frequency wave (envelope). We say that the envelope modulates the amplitude of the carrier. For example, an AM radio station will send out radio waves $V(t) = [A + Bf(t)] \cos(2\pi ct)$, where A and B are constants, c is the carrier frequency, and f(t) is the audio signal that you hear.

5. Use the program you created for exercise 3, page 60, to plot a set of data points. Then draw the best-fit line that runs through the points.

6. For files of data points such as the (x,y) pairs in exercise 5 write a program that plots all values on the screen no matter what their range. Use the procedure "trans" described in Appendix C, after declaring that your data x, y will actually be a pair of "samples". The order of steps in the program should be as follows:

   (a) Fill x and y with data using "getpairs".
   (b) Compute maximum and minimum values for samples x and y using "minpoint" and "maxpoint" functions.
   (c) Compute "xscale", "yscale", "xoff", and "yoff" (see pages 66 and 67 and Appendix C, pages 219 and 220).
   (d) Scale both samples to fit onto the screen by calling the "trans" procedure twice.
   (e) Then "plot" each point x[n], y[n] onto the screen "(n = 1..size)". Every point should now lie within screen limits.

7. Combine the work of exercises 5 and 6 to create a *general best-fit* program that does the following:

   (a) Accepts files of data pairs x[n], y[n]; one pair on each row of the file.
   (b) Scales and plots all data points as in exercise 6.
   (c) Draws the best-fit straight line through the field of points.
   (d) Puts numerical values "bestm" and "bestb" on the display (if your terminal has the capability).
   (e) Computes other numbers of interest, such as mean and error of the x data, for the display.

Answers

1. Axes can be drawn with the following statements:

```
move(−1.0, 0.0);
draw(1.0, 0.0);
move(0.0, 1.0);
draw(0.0, −1.0);
```

Asymptotes can be drawn with the following statements

```
for j := 0 to 1 do begin
      dash(pi * (2 * j + 1)/2);
      dash(−pi * (2 * j + 1)/2);
end;
```

where procedure "dash" is

```
procedure dash(x: real);
      var j: integer;
      begin
            move(x,−1);
            for j := 1 to 20 do begin
                  draw(x, −1 + j/10 − 1/20);
                  move(x, −1 + j/10);
            end;
      end;
```

2. Draw each point having abs(x) > 1.
3. You can do polar problems more easily if you define your own procedures such as

```
procedure polarmove(r, theta: real);
begin
    move(r * cos(theta), r * sin(theta))
end;
procedure polardraw(r, theta: real);
begin
    draw(r * cos(theta), r * sin(theta))
end;
```

For the problem in exercise 3, "h := 0" gives a circle.
4. Your picture should look like that in Figure 5.4.
5. A line of slope m and intercept b is drawn by

```
move(xmin, m * xmin + b); draw(xmax, m * xmax + b)
```

6. These steps are straightforward applications of the library procedures of Appendix C. Remember that maxpoint and minpoint return the positions, not the values, of the extremal data.

7. This problem is a straightforward application of the existing procedures; the final test is the display itself.

PARAMETRIC CURVES

The exercises on pages 68 to 70 included the task of drawing an ellipse in polar coordinates. Often it is possible to express x and y, not as functions of r and theta, but as functions of a *single parameter.* In general, a *parametric curve* is a vector r(t), where t is the parameter and x, y, z, ... are the components of r.

The program "model" in Figure 5.5 plots two-dimensional parametric curves using a procedure "sketch" that modifies the coordinates as well as draws the picture. For any particular graphics task it is important to determine the correct condition for terminating the picture. Notice the "until t > lim" statement in the program "model". This may not be the optimal termination condition, as we shall see in the following exercises.

```pascal
program model(output);
const
     dt = 0.001;
     lim = 1;
var
     x, y, t: real;

     procedure sketch(var u, v: real; t: real);
     begin
         u :=          .     (* whatever function of t you wish. *)
         v :=                (* another function of t *)
         if t = 0 then
             move(u, v)
         else
             draw(u, v)
         end; { sketch }

begin
     t := 0;
     clear;
     graph;
     repeat
       sketch(x, y, t);
       t := t + dt
     until t > lim
end.
```

Figure 5.5

Exercises

1. Modify the program "model" in Figure 5.5 to accept as input the real numbers:

    angle   vel   accel

Define u(t) and v(t) by the relations

u(t) = −1 + vel * cos(angle) * t;
v(t) = −1 + vel * sin(angle) * t − accel * t * t/2;

so that (u,v) is the position of a body thrown with initial speed "vel" at an
initial angle "angle" with respect to the horizontal. The acceleration of gravity
is the number "accel". Perform "cannon-shot" experiments by choosing the
input numbers and graphing the motion of the "cannonball." For given
"accel" and "vel", what is the optimum angle; i.e., what angle gives the
greatest range of the projectile? Remember to figure out the correct
termination condition for a cannon shot. Ask yourself what mathematical
statement is equivalent to "the body has hit the ground."

2. Investigate properties of *Lissajous figures* whose parametric forms are

u = cos(m*t)
v = sin(n*t)

If m, n are integers the picture closes upon itself after a finite time t and
repeats. These pictures appear on an oscilloscope driven horizontally by a
signal u(t) and vertically by a signal v(t) when the ratio of signal frequencies
is m/n.

3. Write a program to draw a parametric spiral defined by u = f(t) cos(t) and
v = f(t) sin(t), where f is an increasing function of time. Try the case
f(t) = 0.1 * exp(t/10), and obtain in this way a *logarithmic spiral*, as appears
in a variety of natural settings, such as the growth of Nautilus seashells and
sunflower seed arrays.

Answers

1. The best angle is 45 degrees.
2. The picture for m = n = 1 is a circle. The picture for m/n = 2 is a
parabola.
3. You should have a spiral on the display. Another function is "f(t) := 0.3
+ t/10", which gives an Archimedean spiral.

THREE-DIMENSIONAL GRAPHICS

The main principle behind the construction of the three-dimensional routines of
Appendix A is that you can specify the appearance of a three-dimensional figure if
you know all of its points (x,y,z) and a fixed set of three angles (call them a, b, c)
that determine the point of view of the observer. The angles a, b, c are defined by
the following operations on a figure:

1. Assume you are looking down the +z axis (toward −z). This is the *null
orientation* (a,b,c) = (0,0,0).
2. Rotate the figure by an angle a around the z axis.

3. Rotate the figure by an angle b around the *new* x axis.
4. Finally, rotate the figure by an angle c around the *new* z axis.

Note that the y axis is not involved directly in these operations, but the set of all possible views is nevertheless obtained. This sequence of operations defines the Euler angles for the view orientation thus achieved. To plot a point, use the statement

splot(x,y,z,a,b,c);

or to draw to a point use the statement

sdraw(x,y,z,a,b,c);

where x, y, and z are the coordinates of the point in its null orientation (see step 1 above).

The constraint on screen size for three-dimensional routines is that each (x,y) must have absolute values not exceeding 1 or else there will be distortion. Remember that the values of (x,y) are changed by rotation, so that sometimes a coordinate initially lying in the interval $(-1,1)$ will, after rotation, lie outside the interval. The easiest way to make sure distortion will not occur is to keep all coordinates in the interval $(-g,g)$, where $g = 1/\sqrt{3}$ or about 0.577.

The routines in Appendix A include options for perspective drawing that take into account the natural distortion caused by relative nearness of the foreground of a figure. These procedures are relevant for later chapters.

Exercises

1. Call the view (a,b,c) = (0,0,0) the "top view." Find the proper angles (a,b,c) for each of the following views:

"top view"	+x *right,* +y *up,* +z *out toward you*
"side view"	+x axis *into* screen, +z *right,* +y *up*
"bottom view"	+x *right,* +y *down,* +z *into* screen
"front view"	+x *right,* +y *into,* +z *up*

Use procedure "axes" in Appendix A to draw the axes lines. Label the axes with small tick marks or letters. Write a program that asks for three Euler angles and draws the axes at the corresponding orientation. You can determine correct values for (a,b,c) in your head, in which case the program is simply to verify your decisions. *Hint:* Note that "bottom view" is obtained for a, c = 0 and a certain b.

2. Write a program that asks for input of a "char", whose value will be "t", "b", "f", "s" and that draws one of the four views (described in exercise 1) of the right parallelepiped of dimensions 0.4 by 0.6 by 0.2. Define a procedure called "drawbox(a,b,c: real)" and call it for the correct set of values (a,b,c). You may like to write a program that displays all four views of the parallelepiped, one in each quadrant of the screen. Use the procedure "rotate(x,y,z,a,b,c)" with which you can rotate coordinates and send them to remote parts of the screen before you draw.

3. Draw a three-dimensional *helicoid*, that is, a curve parameterized according to:

$x = \cos(t) * f;$
$y = \sin(t) * f;$
$z = g * t$

where f and g are constants. Good views are obtained for various $(a,b,c) \neq (0,0,0)$.

4. Write a program to draw the earth at an arbitrary tilt. The program is to readln(a,b,c), and draw lines of latitude and longitude in 15 degree ($\pi/12$) increments each. It is best to use spherical coordinates r,theta,phi defined as follows:

r: radius of sphere, constant for this problem
theta: latitude (0° north, 180° south)
phi: longitude (360° is once around)

Increment theta by $\pi/12$. For each increment let the *azimuth* phi run through 2π in small steps. This will draw the lines of latitude. Longitudinal lines are drawn in a similar way. The required relationships between (x,y,z) and (r,theta,phi) are

$z = r * \cos (\text{theta})$
$x = r * \sin(\text{theta}) * \cos(\text{phi})$
$y = r * \sin(\text{theta}) * \sin(\text{phi})$

For example, the North Pole is attained by

smove(0,0,r,a,b,c);

and the South Pole by

smove(0,0,−r,a,b,c);

A similar program was used to create part of the earth's globe in Figure 5.7.

Answers

1. Some Euler angles for the views are (0, pi, 0), (pi/2, pi/2, −pi/2), (0, −pi/2, 0).
2. First, work out on paper what the drawing sequence should be. Make sure to lift your graphics pen with the "move" procedure whenever you have to start drawing a separate section of the box outline. The x, y, z take on values ±0.4, ±0.6, ±0.2.

3. The problem is solved when you see springs drawn for various sets of Euler angles.

4. Here is a section from a working program:

```
{plot north and south poles}
splot(0, 0, r, a, b, c);
splot(0, 0, −r, a, b, c);
{draw lines of latitude}
theta := pi / 12; {pi/12 radians = 15 degrees}
repeat
    phi := 0
    smove(r * sin(theta), 0, r * cos(theta), a, b, c);
    z := r * cos(theta);
    repeat
        phi := phi + delta;
        x := r * cos(phi) * sin(theta);
        y := r * sin(phi) * sin(theta);
        splot(x, y, z, a, b, c)
    until phi >= 2 * pi;
    theta:= theta + pi/12;
until theta > 11/12 * pi;
```

The lines of longitude can be similarly plotted.

VERSATILITY OF GRAPHICS

Graphics is obviously a versatile way to provide meaningful output. At times, however, this very versatility may make it difficult to write programs. Should you use a surface or a series of histograms or a knot of trajectories or a contour map or some other graphic form?

One way to determine optimal output format is to gain experience. You need to have had enough feedback, in the form of successful and unsuccessful graphics displays, to be able to anticipate the usefulness of the intended output. This chapter will therefore end with problems that have been carefully selected both to be typical and to cover a broad range of output styles.

Figures 5.6 to 5.13 illustrate graphics output styles for various research problems. Look these over before doing the following exercises.

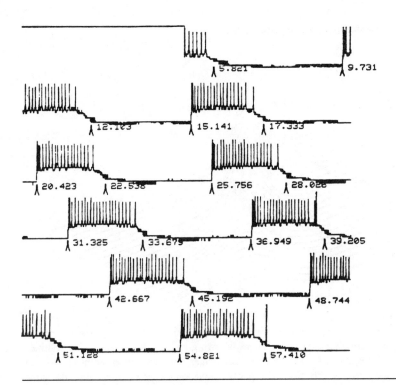

Figure 5.6 Nerve impulses from a single cell of the barnacle *Balanus cariosus,* representing 60 seconds of electrical activity. The display and the timing markers were created using Pascal programs such as those discussed in Chapter 9, the section titled *Biological Signal Processing*. The data are courtesy of G. F. Gwilliam.

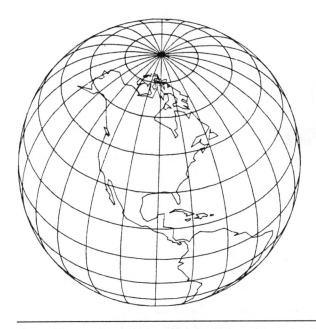

Figure 5.7 Three-dimensional plot of earth using the "plib3.i" three-dimensional Pascal routines (Appendix A). The task of drawing latitude and longitude is not too difficult, but the tasks of hiding lines and creating the continents require some real expertise. This display was created by Robert Henley.

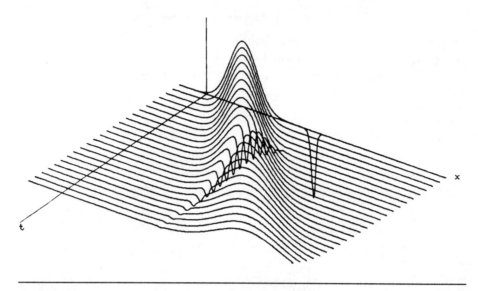

Figure 5.8 A reflectionless scattering event, showing space-time format using routines from Chapter 8 and Appendix E. The wave packet strikes a special potential through which all energy must pass with no reflection. The packet exits somewhat squat but with the same area with which it entered. Display courtesy of Barbara Litt. See also Figure 8.16.

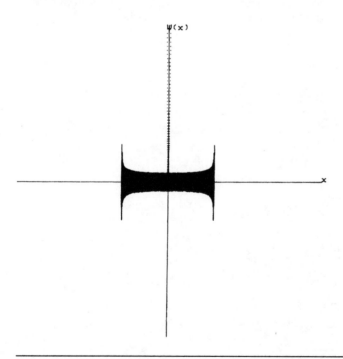

Figure 5.9 Graph of the 8000th excited state of the quantum oscillator. The curve has 8000 zero crossings. The classical limit distribution, a bowl-shaped envelope to the graph, is easily seen. Quantum mechanics using Pascal is discussed in Chapters 7 and 8.

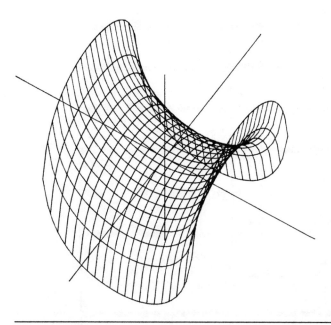

Figure 5.10 This minimal surface is the shape that a soap film will take when connecting the indicated wire boundary. No other curve connects this boundary with less total area. The plot is a direct graph of the minimal surface $z = \ln[\cos(x)] - \ln[\cos(y)]$.

Exercises

These exercises have been designed to enhance and test your versatility with graphics and programming techniques covered in Chapters 1 to 5. In many cases there is no best solution, but you should always try to find an *efficient* one. Be sure to think through the problem before writing any of the program and, above all, do not simply fiddle with the graphics terminal. Examples of research-oriented graphics are included in Chapters 6 to 9.

1. Write a program that accepts input of a positive integer n (greater than 2) and draws a regular n-gon.
2. Write a program to draw the curve

$$|x|^a + |y|^a = 1$$

where a is a positive real constant. What does the curve look like when a is extremely large? What figures do you get for the special cases a = 1, a = 2, a is nearly 0?
3. Write a program that models the two-dimensional motion of a single ideal gas molecule bouncing around inside a square box. Guess what the condition will be on the initial launch angle (with respect to the horizontal box sides) such that the path will eventually repeat itself.
4. Let a particle move such that its speed is constant but its velocity vector always makes a fixed angle with its radius vector. Write a program that

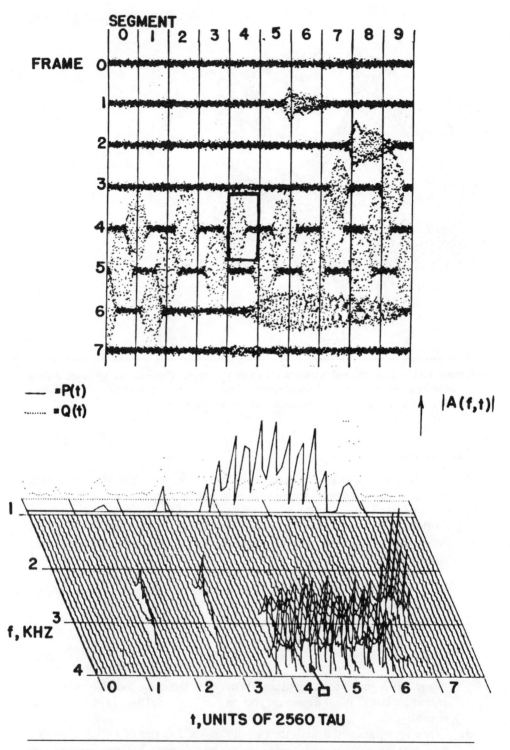

SEGMENT
0 1 2 3 4 5 6 7 8 9

FRAME 0
1
2
3
4
5
6
7

—— = P(t)
········ = Q(t)

↑ |A(f,t)|

f, KHZ

1
2
3
4

0 1 2 3 4 5 6 7

t, UNITS OF 2560 TAU

Figure 5.11 The upper figure is a 1-second stretch of *Melospiza melodia* (song sparrow) bird song. Fast Fourier transforms were made on this data (by S. Cochran). The running spectrum is shown in the bottom figure. Such signal processing is discussed in Chapter 6. Fast Fourier transforms are discussed in Chapter 7.

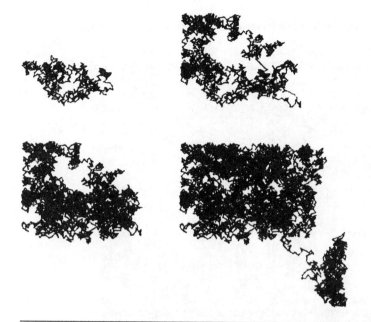

Figure 5.12 Successive stages in the life of a particle in Brownian motion. This sequence is a good example of simple graphics modeling of the physical world and the use of random real numbers. If the particle were a nitrogen molecule, this is how it might move in a large auditorium box at room temperature with several seconds having elapsed in the last frame.

draws such trajectories for various fixed angles u. When $u = \pi/2$, you should get a circle. Otherwise you will get *logarithmic spirals,* as mentioned in exercise 3, page 73.

5. Model *Brownian motion* as follows. When at (x,y), the next point to which you draw has coordinates

```
x := x + dx * (2 * random(1) − 1);
y := y + dy * (2 * random(1) − 1);
```

where dx = dy is a small real value. This is a model of the continuous analog to the random walk concepts discussed in Chapter 4. Intricate paths are generated, modeling the motion of for example a small smoke particle or plastic microbead as it is jiggled by air molecules.

6. By iteration, graph the solution of the *damped harmonic oscillator* motion x(t) satisfying

$$m \frac{d^2x}{dt^2} + a \frac{dx}{dt} + kx = 0$$

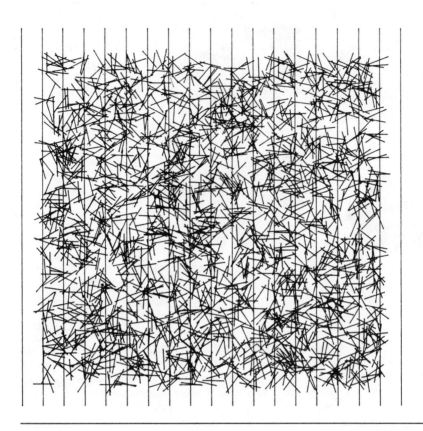

Figure 5.13 Buffon's needle experiment, pi = 3.13e + 00. See Chapter 4, exercise 6, page 57.

where m, a, and k are all positive constants. Let the horizontal axis be t and the vertical axis be x(t); start with initial conditions

$$x(0) = 1 \qquad \frac{dx}{dt}(0) = (0)$$

Verify the following theoretical predictions for such an oscillator:
(a) As a approaches zero, the oscillation frequency approaches

$$\frac{\text{sqrt}(k/m)}{2\pi}$$

(b) The *damping time* [time required for the amplitude to fall off to exp(−1)] is given by T = 2m/a.

7. Write a program to model the focusing of parallel rays by a parabolic mirror onto a single focal point.

8. Consider the points $(-1,1)$ and $(0,-1)$ on the graphics screen. A straight ray is to leave $(-1,1)$ and strike the x axis. At this junction it is to travel in a straight path to the point $(0,-1)$. Assume, however, that the speed of the ray is v1 for positive y and v2 for negative y. The idea is to find where the x axis is to be struck if you invoke *Fermat's principle,* which states that the total transit time be a minimum. What can you say about the path of least time in general?

   This problem gives a model for refraction of light rays at the interface of two media (such as air and glass). Various angles at the interface (x axis) will satisfy *Snell's law of refraction*, which can be found in most elementary physics texts. This sort of program can be used to design lenses and other optical devices.

9. Write a program to model a bouncing ball that loses 10% of its kinetic energy every time it strikes the floor. Does the ball come to rest in a finite time? (Be careful with this last question.)

10. Write a program that draws a three-dimensional cone at an arbitrary orientation.

11. Write a program that draws drafting views of a regular tetrahedron (see Chapter 3, exercise 1, page 44).

Answers

1. The r coordinate is to be a constant, for example 0.8, and the polar angle moves in steps of pi/n. Then x, y are given by

   ```
   x := r * cos(angle)
   y := r * sin(angle)
   ```

   A simple "for j := 1 to n" statement will draw the n-gon.

2. For large a, the curve is a square; for a nearly 0 it is a four-pointed starfish; for a = 2, it is a circle. For a = 1 it is a square.

3. The problem of finding the point on a struck wall can be solved algebraically for arbitrary initial point and ray angle, but this is complicated. A slower approach that is easier to comprehend and program is to start from any point and draw small increments in the appropriate direction, until either abs(x) > 1 or abs(y) > 1, meaning the respective wall has been touched. After touching, you negate the associated drawing increment coordinate, causing reflection, and continue to draw. The process starts out with initial (x0,y0) and initial

   ```
   dx := dr * cos(theta);
   dy := dr * sin(theta);
   ```

   where dr is a suitably small constant (length of each small drawing segment). You draw with a repeat loop until either abs(x) > 1 or abs(y) > 1, indicating a wall is struck. You then do "dx := −dx" for vertical-wall strike or "dy := −dy" for horizontal-wall strike. This causes reflection; you continue to draw the small increments. The theoretical result is that if the initial angle has a rational tangent, then the path repeats itself after a finite number of reflections.

4. The spirals are similar to the exponential type in Chapter 3, exercise 3, page 73. You draw, from the point

$(r * \cos(theta), r * \sin(theta))$

by a small increment given by

dx := dr * cos(theta + u);
dy := dr * sin(theta + u);

where dr is a suitably small constant. After drawing to this nearby point, you must update the value of r.

5. The pictures should look like the Brownian motion figure (Figure 5.12). Reflection off a wall is performed by negating dx or dy as the x or y coordinate, respectively, tries to jump through a wall [walls are defined by $\text{abs}(x) = 1$ or $\text{abs}(y) = 1$].

6. You use the program "sho" in Chapter 3, Figure 3.7, with the appropriate modification for the damping term a*dx / dt, and plot the points x(t) (vertical axis) versus t (horizontal axis). The picture should be a sine-wave oscillation that damps, i.e., pinches off toward the right. It should fall to 37% of initial amplitude at the time $t = 2m/a$.

7. This can be solved using the discussion for exercise 3 above.

8. Let a and b be the angles the solution trajectory makes with the vertical axis in regions $y > 0$ and $y < 0$, respectively. Snell's law dictates that $\sin(a)/\sin(b) = v1/v2$, and this should be true for your display. When v1 is chosen equal to v2, there must not be any bend at the $y = 0$ interface.

9. The ball comes to rest in a finite time because the sum of all bounce times is a convergent series. A perfect ball with 90% restitution, as given, will bounce faster and faster, its bounce rate will go to infinity, but it will come to rest in a finite time.

10. A cone can be drawn by the set of operations

smove(x,y,0,a,b,c);
sdraw(0,0,0.7,a,b,c);

where Euler angles (a,b,c) are predetermined, and points (x,y) lie on a circle.

11. This is a straightforward generalization of Chapter 3, exercise 1, page 44. Typical views are "top", "side", "front", and possibly an oblique view.

6 | Examples from Mathematics

COMPLEX NUMBERS

In Pascal programming, it is important to give your functions and types good *names*. Advanced programs are difficult to debug, and it helps if the program is clear and the names have meaning.

The reals and integers are not the only mathematical entities we can manipulate within programs. In Chapters 3 and 4, for example, we used arrays in calculus and statistics. Arrays also have relevance to computer solutions. An example of the meaningful use of arrays is the declaration of a *complex number*.

One way to declare a complex number is as a vector of two reals:

```
const component = (re,im);
type complex = array[component] of real;
```

The declaration

```
var x: complex;
```

allows you to set up the complex number $3 + 2i$ by

```
x[re] := 3; x[im] := 2;
```

and so on.

Another way to declare complex numbers is as follows. The Pascal *record* is declared as

```
type complex = record
               re, im: real
               end;
```

The declaration

var x: complex;

allows you to set up the complex number $3 + 2i$ as

x.re := 3; x.im := 2;

which is easier to work with than the array notation. Also, complicated structures can be created in the form of records, especially *variant records.* Details on record types are in Chapter 1 references.

Since functions cannot be made to return records, you usually multiply, divide, add, and subtract using procedures. An example is the calculation of the *Cauchy integral:*

$$\oint f(z) \; dz$$

The program "residue" in Figure 6.1 shows how records can be used to calculate such integrals. It performs the integral

$$I = \oint \frac{dz}{z - y}$$

where the contour is the unit square in the complex plane. In the program, y is set to $0.1 + 0i$ so that by the residue theorem of complex analysis, the integral depends only on the pole at $z = y$ and thus has the value

$$I = 2\pi i$$

This result printed out by the program "residue" is the number pair 0, 6.28.

Another example of manipulation of complex numbers is the calculation of *analytic functions.* Consider the *gamma function* with the Euler product form

$$\frac{1}{\Gamma(z)} = z \; \exp(\gamma z) \; \prod_{n=1}^{\infty} \left(1 + \frac{z}{n} \right) \exp\left(\frac{-z}{n} \right)$$

where γ is Euler's constant (see exercise 2, page 20). For positive integers n, $\Gamma(n) = (n - 1)!$, so the function is a generalized factorial function. The program "gamma" in Figure 6.2 calculates complex gamma function values. Notice the new procedure "cexp", which exponentiates the complex number a and places the result in var c.

```
program residue(input, output);
type
    complex =
        record
            re, im: real
        end;
var
    c, y, z, integral, dz: complex;
    u, v: integer;

    procedure mul(a, b: complex; var c: complex);
(* complex multiply c = a*b *)
    begin
        c.re := a.re * b.re - a.im * b.im;
        c.im := a.re * b.im + a.im * b.re
    end; { mul }

    procedure add(a, b: complex; var c: complex);
(* complex add c = a+b *)
    begin
        c.re := a.re + b.re;
        c.im := a.im + b.im
    end; { add }

    function cmod(x: complex): real;
(* modulus of complex x *)
    begin
        cmod := sqrt(sqr(x.re) + sqr(x.im))
    end; { cmod }

    procedure invert(a: complex; var c: complex);
(* complex reciprocate c = 1/a *)
    var
        den: real;
    begin
        den := sqr(cmod(a));
        c.im := -(a.im / den);
        c.re := a.re / den
    end; { invert }

begin
    z.re := 1;
    z.im := 1;              (* start at z=1+i *)
    y.re := -0.1;
    y.im := 0;
    for u := 0 to 3 do begin
        dz.re := 0;
        dz.im := 0;
        if u mod 2 = 0 then
            dz.re := 0.002 * (u - 1)
        else
            dz.im := 0.002 * (u - 2);
        (* we now have dz running along a side of square *)
        for v := 1 to 1000 do begin
            add(z, y, c);           (* get z-0.1 in c *)
            invert(c, c);           (* get c = 1/(z-0.1) *)
            mul(c, dz, c);          (* get c as integrand *)
            add(integral, c, c);
            integral := c;
            add(z, dz, c);
            z := c
        end
    end;
    writeln(integral.re, integral.im)
end.
```

Figure 6.1

```pascal
program gamma(input, output);
const
    gam = 0.5772156649015328606065l2;
    bign = 100;
type
    complex =
        record
            re, im: real
        end;
var
    temp, temp2, prod, c, x, y, z: complex;
    n: integer;

    procedure mul(a, b: complex; var c: complex);
(* complex multiply c = a*b *)
    begin
        c.re := a.re * b.re - a.im * b.im;
        c.im := a.re * b.im + a.im * b.re
    end; { mul }

    procedure add(a, b: complex; var c: complex);
(* complex add c = a+b *)
    begin
        c.re := a.re + b.re;
        c.im := a.im + b.im
    end; { add }

    function cmod(x: complex): real;
(* modulus of complex x *)
    begin
        cmod := sqrt(sqr(x.re) + sqr(x.im))
    end; { cmod }

    procedure invert(a: complex; var c: complex);
(* complex reciprocate c = 1/a *)
    var
        den: real;
    begin
        den := sqr(cmod(a));
        c.im := -(a.im / den);
        c.re := a.re / den
    end; { invert }

    procedure cexp(a: complex; var c: complex);
    var
        temp: real;
    begin
        temp := exp(a.re);
        c.re := temp * cos(a.im);
        c.im := temp * sin(a.im)
    end; { cexp }

    begin
        repeat
            readln(z.re, z.im);
            x.re := 1;
            x.im := 0;
            prod := x;
            (* begin constructing Euler's product *)
            for n := 1 to bign do begin
```

Figure 6.2

```
        y.re := 1 / n;
        y.im := 0;
        mul(y, z, c);
        temp := c;
        cexp(c, c);
        invert(c, c);
        temp2 := c;
        add(x, temp, c);
        mul(c, temp2, c);
        mul(c, prod, c);
        prod := c
      end;
      x.re := gam;
      mul(x, z, c);
      cexp(c, c);
      mul(prod, c, c);
      mul(c, z, c);
      invert(c, c);
      writeln(c.re, c.im)
    until 0 = 1
end.
```

Figure 6.2 (continued)

Exercises

1. Modify the program "gamma" in Figure 6.2 to take into account the known reduction and recurrence relations for the gamma function. For example, $\Gamma(z + 1) = z\,\Gamma(z)$, so that the error that now occurs when "gamma" is run for large real z can be eliminated by first reducing the argument.

2. Is it generally faster to do contour integrals such as in the program "residue" or to seek out all poles of a function and add up residues? The two approaches can be compared, for example, in the evaluation of $\oint(16z^4 - 1)^{-1}\,dz$, the contour being the unit circle.

3. Write a program that finds *all* the roots of a polynomial. First, consider that the *fundamental theorem of algebra* says that an nth degree polynomial with complex coefficients will have at most n complex roots. Therefore, you should investigate all the deeper properties of the solutions using abstract mathematics before writing code. For example, the complex roots will generally appear in pairs a + bi and a − bi, and you do not want to use computer time finding all the roots you could find with algebra. Consult a numerical analysis book for known algorithms for finding such roots.

4. Write a program that investigates some property of *Gaussian integers* a + bi, where a and b are real integers. For example, can you factor a + bi? If so, how? Consider that 2 is a real prime but not a Gaussian prime since $2 = (1 + i)(1 - i)$.

5. Write a program to compute the functions cos(x) and sin(x) using the formula

$$\cos x + i \sin x = \lim_{n \to \infty} \left(1 + \frac{ix}{n}\right)^n$$

Test your work for various x's by comparing results with the standard Pascal cosine and sine functions.

Answers

1. The relevant reduction techniques are in the gamma function portion of "plibl.i" in Appendix D.
2. The method of seeking poles is faster if you know that all singularities lie in some bounded region.
3. A good way to test your program is to enter polynomials prefactored in the form $(x - r1) * (x - r2) * \cdots$.
4. The number $a + bi$, if factorable, has decomposition $(c + di) * (e + fi)$ so that you need to solve

   $a = ce - df$
   $b = cf + de$

   But many cases can be eliminated by noting that if $sqr(a) + sqr(b)$ is a prime, you cannot factor $a + bi$ because this would mean $(sqr(c) + sqr(d))$ and $(sqr(e) + sqr(f))$ were both factors of that prime. Also note that factorization is only defined for the gaussian integers up to ambiguous *units*, for example, $1 = (-i) * i$, so that you stop factorization when any further factors have absolute value equal to 1.
5. There are two good approaches to this problem. One is to write a loop that generates the product $(1 + ix/n)(1 + ix/n) \cdots$ and then outputs $\cos (x)$, $\sin (x)$ as the real, imaginery parts, respectively. The other approach is to define real and imaginary parts of both "cln" (complex log) and "cexp" (complex exponential) functions. Then $\cos x + i \sin x = \lim_{n \to \infty}$ $cexp(n * cln(1 + ix/n))$. *Hint:* For general complex number $z = a + bi$ and $a > 0$, $cln(z) = \ln \sqrt{a^2 + b^2} + i \arctan (b/a)$.

FOURIER SERIES

Let signal[n] denote a collection of complex numbers corresponding to a time series. The index n is an integer, and *real time* t is related to that index by

$t = n * tau$

where tau is the time between samples, or *sampling time*. What is often called the *sampling rate* is just tau^{-1}. The program "generator" in Figure 6.3 creates a certain signal. The function being generated is

$$signal[n] = \sum_{j=1, 2, 3, 4} \sin(2\pi n * tau * f_j)$$

```
program generator (output);
(* create facsimile data for four tones rising *)
const
     tau = 0.0001172; (* time between samples *)
     pi = 3.14159265359;
     totaln = 10240; (* total number of data *)
var
     n, u: integer;
     signal: real;
     f: array [0..3] of real;
begin
     for n := 1 to totaln do begin
         signal := 0;
         for u := 0 to 3 do begin
             f[u] := 600 + 100 * u;
             f[u] := f[u] * (1 + n / 10000);
             signal := sin(2 * pi * f[u] * n * tau) + signal
         end;
         writeln(signal * 0.2: 5: 5)
     end
end.
```

Figure 6.3

where the four frequencies $f_1, \ldots, f_4$ are slowly increasing in time. Thus the signal is a sum of four *gliding* pure notes. A plot of this signal in the form of an *oscillogram* is shown in Figure 6.4. The signal is displayed in four *frames* so that the time axis wraps around three times. It is not immediately clear how to recover the frequency information given the signal itself. We would like to have a technique

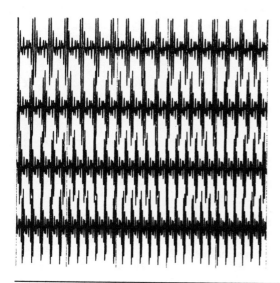

Figure 6.4

```
program dft(input, output);
(* compute strightforward DFT for input data file *)
const
    pi = 3.1415926536;
    maxsize = 256;        (* size of each block to be DFT'd *)
    tau = 0.0001172;      (* time between samples *)
    peak = 1;             (* scaling factor - largest absolute datum *)
var
    sn, cn: array [0..630] of real;
    num: integer;
    s, c, t, el, b, yy: real;
    factor: real;
    fmin, fmax, df, w: integer;
    y: array [1..maxsize] of real;
    om: real;
    arg: real;
    ctr: integer;
    freq: integer;
    temp: real;
begin
    for ctr := 1 to 630 do begin
        sn[ctr] := sin(ctr / 100);
        cn[ctr] := cos(ctr / 100)
    end;
    b := 2 * pi;
    readln(fmin, fmax, df);
    factor := 0.5 * maxsize * peak;
    repeat
        freq := fmin - df;
        num := 0;
        repeat
            num := num + 1;
            readln(y[num])
        until num = maxsize;
        repeat
            freq := freq + df;
            om := b * freq;
            s := 0;
            c := 0;
            t := 0;
            arg := 0;
            el := om * tau;
            for ctr := 1 to num do begin
                arg := arg + el;
                if arg > b then
                    arg := arg - b;
                w := trunc(100 * arg);
                yy := y[ctr];
                s := s + yy * sn[w];
                c := c + yy * cn[w]
            end;
            temp := sqr(s) + sqr(c);
            temp := sqrt(temp) / factor;
            writeln(freq, temp: 3: 3)
        until freq > fmax - df;
        writeln
    until eof
end.
```

Figure 6.5

for estimating the *gliding frequencies* of the above signal. Recovery of frequency information from a signal is a good application for *Fourier analysis.*

For a stream of data called "signal", we define the *Fourier transform,* or *spectrum,* to be a function q defined for real m by

$$q(m) = \sum_{n=0}^{N-1} \exp\left(\frac{2\pi imn}{N}\right) * \text{signal}[n]$$

where we assume there are a total of N signal values, signal [0] through signal [N − 1]. The program "dft" in Figure 6.5 computes the Fourier transform in a straightforward way that does not use complex numbers (but could be modified to do so). Think of q(m) as the amount of signal having a certain frequency.

One of the confusing aspects of Fourier transform programming is the definition of *frequency*. The correct formula for frequency f is

$$f = \frac{m}{N * tau}$$

where tau is the time between signal values. The "dft" program gives the modulus q(m) versus various frequencies running from "fmin" to "fmax" in steps of df. These last three parameters are input at run time. Note that tau does not appear in the theoretical definition of the spectrum function q. This is because frequency depends on the time scale used, and this time scale is embodied in the number tau. Note that tau is declared as a constant in the program "dft". Tau is needed for proper scaling for output because we input frequency instead of m. *Question:* Where is m in the program "dft"? *Answer:* It is not exhibited directly but is equal to maxsize * freq * tau on the basis of the previous definition for f. Also, the var "ctr" in the program "dft" is the running summand index n in the definition of q(m).

Several mathematical techniques were used to write the program "dft". First, arrays "sn" and "cn" were filled to optimize the speed of computation. This is called the method of generating a *lookup table.* Indeed, the main loop (starting "for ctr := 1 to num") does not involve calculation of "sin" and "cos" directly; rather these values are looked up in the arrays "sn" and "cn". Second, defining the auxiliary variables "yy", "om", "el", and so on, helps optimize speed because you eliminate the time it would take to evaluate their respective expressions if they were not declared. For example, "yy" allows you to look up y[ctr] only once in the loop pass.

The output of the program will be a table of frequency versus the function q. An example of table output is shown later in Figure 6.8. Graphics output of the spectrum q is shown now in Figure 6.6.

Figure 6.6 shows the graphics output of the frequency spectrum for each of the four frames shown in the oscillogram in Figure 6.4. There are some interesting effects, notably the lack of sharpness of the frequency peaks and the slow glide upward of each frequency component. The latter effect is inherent in the output of the program "generator" that made the signal, but the former effect is an *uncertainty principle,* which always holds. Frequencies are never precisely defined unless the number of signal data is infinite.

The problem with the "dft" program, which calculates the straightforward complex transform, is that it typically runs very slowly. In the late 1960s people began to realize that there is a lot of redundancy in the computation of the transform as a straight sum because many of the factors exp (· · ·) are repetitive. The *fast Fourier transform* (FFT) was thus created (Cooley and Tukey, 1965; Ralston and Rabinowitz, 1978).

Some nomenclature is appropriate here: DFT stands for *discrete Fourier transform* and is sometimes taken to mean the function q(m) for integers m = 0,

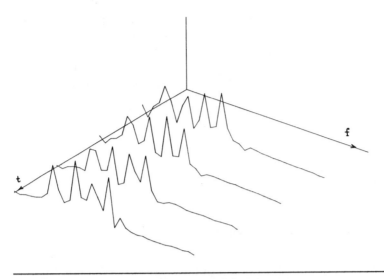

Figure 6.6 "DFT" output created by program "dft" showing spectral peaks for each of four frequencies. The peaks glide (in f direction) with time t.

1, ..., N − 1. The FFT computes the discrete Fourier transform in this integer case. In fact, you can think of the FFT as an *array* q[m] (note square brackets) to be computed by the modern techniques that exploit redundancy of the exp( ) factors.

The program "fft" in Figure 6.7 computes the transform of input signal files. The arrays xr[ ] and xi[ ] are the real and imaginary parts, respectively, of the spectrum array q[ ], since the algorithm simply replaces data with spectrum.

For a total of N elements in the array "signal", the time it takes to do the DFT is proportional to N^2, whereas the time it takes to do the FFT is proportional to N log N. The FFT is much faster, even on the signal example we have chosen. The final output from the two Fourier programs in Figures 6.5 and 6.7 is shown in Figure 6.8. The first column is the output from the program "dft" in Figure 6.5, and the second column is the output from the program "fft" in Figure 6.7. When output FFT is graphed, it appears identical to the DFT graph (see Figure 6.6). The only significant difference between the two types of spectrum output is that DFT frequencies (Figure 6.8, column beginning with 500) were determined at run time to be in steps of df = 33, whereas the FFT algorithm chooses its own step size on the basis of the constraint that the spectrum array q have integer indices.

The program "fft" requires an input file having a parameter string as its first line. This is described in the opening comment of the "fft" listing (Figure 6.7). The numbers used for the DFT-FFT comparison table in Figure 6.8 are

1024 256 0.0001172 500 1500

which allows transformation of 256 lines of output from the program "generator" (Figure 6.3).

For *spectrograms*, which you can think of as Pascal-generated equivalents of spectrum analyzer or spectrograph output, you can combine the following:

```
(* This program computes running FFT spectrograms, for arbitrary
   integer sample sizes.  The input file is to contain one line
   of parameters, followed by (one datum per line) signal
   numbers.
   The overall format is:

        s n tau fmin fmax (this is 1st line of input file)
        xr[1]             (first signal datum)
        xr[2]             (second signal datum)
   etc.

   Where the parameters are:
        s: total number of data samples in input file
        n: number of samples per spectrum
        tau: actual real-time increment so that frequencies are in Hz.
             (that is, sampling rate is 1/tau)
        fmin, fmax:  frequency limits for which you desire spectral data

   Examples of usage:
1) you wish to fft 1024 samples taken every 0.1 seconds.  The
   input file may begin with the line:
   1024 1024 0.1 100 200
   which will give frequency-amplitude output from 100 to 200 Hz.,
   followed by 1024 data, each on their own line.
2) Same as (1) but you want four spectra, each performed for 256
   data values.  The data are the same, but the first input line is:
   1024 256 0.1 100 200

Output is absolute amplitude vs. frequency.  Power would be the
amplitude squared, all of this derived from real and imaginary
spectrum returned in arrays xr[], xi[] respectively.
*)

program fft(input, output);
const
    pi = 3.1415926535897932;
    peak = 1;   (* maximum data value *)
var
    s, ctr, p, p0, i, j, n, a, b, c, l, k, m: integer;
    factor, df, z, temp: real;
    sum, car, q, arg: integer;
    x, y, tau, fmin, fmax: real;
    aa, pr: array [1..20] of integer;
    cc: array [0..19] of integer;
    xr, xi, tr, ti, zz, cn, sn: array [0..511] of real; (*zz == zeroes*)
    jj: array [0..511] of integer;

begin
    readln(x, y, tau, fmin, fmax);
        (* you can set these five args to be hard values *)
    s := trunc(x);
    n := trunc(y);
    factor := 1 / (0.5 * n * peak);
    df := 1 / (n * tau);
        (*now factor n*)
    m := n;
    l := 0;
    j := 2;
        (*pr is array of prime factors of n*)
    repeat
        while m mod j = 0 do begin
            m := m div j;
            l := l + 1;
            pr[l] := j
        end;
        j := j + 2;
        if j = 4 then
            j := 3
```

Figure 6.7

```
        until m = 1;
            (*re-order array in reverse-complement style *)
        cc[0] := 1;
        for i := 1 to 1 - 1 do
            cc[i] := cc[i - 1] * pr[i];
        for m := 1 to 20 do
            aa[m] := 0;
        jj[0] := 0;
            (*jj array is input order*)
        for i := 1 to n - 1 do begin
            j := 1;
            car := 1;
            repeat
                aa[j] := aa[j] + car;
                car := aa[j] div pr[j];
                aa[j] := aa[j] mod pr[j];
                j := j - 1
            until car = 0;
            sum := 0;
            for q := 0 to 1 - 1 do
                sum := sum + aa[q + 1] * cc[q];
            jj[sum] := i
        end;
            (*create sin, cos arrays*)
        for k := 0 to n - 1 do begin
            z := 2 * pi * k / n;
            sn[k] := sin(z);
            cn[k] := cos(z)
        end;
        for ctr := 1 to s div n do begin
            for i := 0 to n - 1 do
                    readln(xr[jj[i]]);
            xi := zz; (* we assume all real input data *)
            c := 1;
            a := 1;
            b := 1;
            repeat
                    a := a * pr[c];
                    for k := 0 to n - 1 do begin
                        arg := a * (k div a) + k mod b;
                        p0 := k * n div a;
                        p := 0;
                        x := 0;
                        y := 0;
                        for q := 0 to pr[c] - 1 do begin
                            x := x + xr[arg] * cn[p] - xi[arg] * sn[p];
                            y := y + xr[arg] * sn[p] + xi[arg] * cn[p];
                            p := (p + p0) mod n;
                            arg := arg + b
                        end;
                        tr[k] := x;
                        ti[k] := y
                    end;
                    xr := tr;
                    xi := ti;
                    c := c - 1;
                    b := a
            until c = 0;
                    (*output sequence*)
            z := 0;
            for k := 0 to n - 1 do begin
                if (z >= fmin) and (z <= fmax) then
                    writeln(z: 3: 3, factor*sqrt(sqr(xr[k])+sqr(xi[k])): 5: 5);
                z := z + df
            end
        end
end.
```

Figure 6.7 (continued)

f(k)	DFT	q(k)
500		0.042
533		0.052
566		0.075
599		0.153
632		0.102
665		0.018
698		0.095
731		0.143
764		0.018
797		0.059
830		0.177
863		0.046
896		0.057
929		0.204
962		0.067
995		0.034
1028		0.024
1061		0.019
1094		0.018
1127		0.015
1160		0.011
1193		0.010
1226		0.008
1259		0.008
1292		0.007
1325		0.006
1358		0.005
1391		0.005
1424		0.004
1457		0.004
1490		0.003

f(k)	FFT	q(k)
533.276		0.05223
566.606		0.07583
599.936		0.16016
633.266		0.09489
666.596		0.02864
699.925		0.11059
733.255		0.12793
766.585		0.01208
799.915		0.07666
833.244		0.15796
866.574		0.03559
899.904		0.05957
933.234		0.19186
966.564		0.07702
999.893		0.04554
1033.223		0.03426
1066.553		0.02812
1099.883		0.02406
1133.212		0.02112
1166.542		0.01887
1199.872		0.01707
1233.202		0.01559
1266.532		0.01436
1299.861		0.01332
1333.191		0.01242
1366.521		0.01164
1399.851		0.01095
1433.180		0.01034
1466.510		0.00980
1499.840		0.00931

Figure 6.8

1. Fourier transform concepts
2. Three-dimensional graphics (Appendix A and Chapter 5)
3. Picture framing (Appendix E and Chapter 8)

Figure 6.9 is a spectrogram plot for 1 second of bird song (song sparrow) that was digitized with a microprocessor and then fed into the "fft" program of Figure 6.7. Chapter 9 discusses other kinds of spectrum analysis.

Exercises

1. Write a program that finds the Fourier transform of the "signal"

   $x[k] = k \bmod 20$

   where the index k goes from 0 to 99. This is a *ramp wave*, which is a series of up-going ramps. What is the derivative with respect to time (this will be the set of differences $x[k + 1] - x[k]$ in the difference approximation)? How does this derivative relate to the transform coefficients q[0], . . . , q[99] for the ramp?

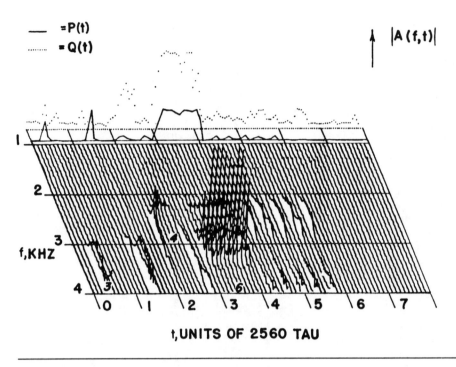

Figure 6.9 Song-sparrow chirp, FFT of 1 second of data, computed by program "fft" (Figure 6.7). See Cochran (1980).

---

2. For a *noise signal* x[k] = 2 * random(1) − 1, what do you expect to be true of the transform coefficients q[k]? Try some experimental programs that work out statistical parameters for the q[k].

3. An interesting and mathematically complicated signal function is the *frequency-modulated* wave

$$x[k] = \exp[i(ak + b \cos ck)]$$

where a, b, and c are real constants. Such a signal has a frequency that changes sinusoidally in time. You might guess that the spectrum would be some sort of distribution lying between the limits of the frequency excursion. Show that this is not so and that, in fact, the spectrum has a complicated structure consisting of a sharp central peak and many *sidebands*. The relative strengths and positions of these sidebands depend strongly on the values of a, b, and c. The proper theoretical analysis of the spectrum involves the *Bessel functions* described later in this chapter. You may be interested in the more general problem encountered by replacement of the cosine function with an arbitrary function f(k), corresponding to the sending of FM carrier by a radio station.

Answers

1. In general, if q(m) is the signal's transform, then the transform of the derivative is roughly proportional to m * q(m). A better formula is

$$Q(m) = q(m) * \left[1 - \exp\left(-2 * pi * i * \frac{m}{N} \right) \right]$$

where Q is the transform of the derivative. Thus the Q = constant * m * q(m) rule is best at low-frequency indices m.

2. The signal is meant to model *white noise,* whose q[k] are also noisy, with roughly equidistributed values across the allowed spectrum.

3. This example is exploratory. One interesting question is as follows. If an FM voltage is $V(t) = \exp[i(at + b \cos ct)]$, what is the instantaneous frequency at time t? *Hint:* Differentiate the phase with respect to time. There should be a peak in the Fourier spectrum at the *average* value of the instantaneous frequency. Given this fact, how do sideband positions depend on a, b, c? How does this in turn dictate how close FM stations can be on the radio dial (88–108 megaHertz).

LINEAR EQUATIONS

If you have n equations in n unknowns arranged as follows:

$$a_{11}x_1 + a_{12}x_2 + \cdots + a_{1n}x_n = c_1$$

$$a_{21}x_1 + \cdots \qquad + a_{2n}x_n = c_2$$

.
.
.

$$a_{n1}x_1 + \cdots \qquad + a_{nn}x_n = c_n$$

it is possible to find a simultaneous solution $(x_1, \ldots, x_n)$, if there is one, by using Cramer's rule. This is made relatively easy by the existing matrix package (Appendix B). The program "linear" in Figure 6.10 solves for x_j in terms of the $n \times n$ matrix (a_{ij}) and the n column (c_j).

If you run the program "linear" and give the coefficients as

```
1  1  1  1  1  2
1  1  1  1  2  2
1  1  1  2  3  3
1  2  3  4  5  6
1  2  3  2  1  2
1  1  2  3  4  4
```

and the column c_j as

```
0  9  8  7  9  8
```

```
program linear(input, output);
(* solve n equations in n unknows, where n <= 16 *)
const
     dim = 16;
type
     matrix = array [1..dim, 1..dim] of real;
     vector = array [1..dim] of real;
var
     m, z: matrix;
     den: real;
     v: vector;
     i, n: integer;
#include "plibm.i"
(* include the matrix library *)
begin
     write('number of variables: ');
     readln(n);
     writeln('enter coefficients: ');
     readmat(n, n, m);
     writeln('enter constants on one line: ');
     readvec(n, v);
     den := det(n, m);
     for i := 1 to n do begin
         z := m;
         changecol(n, n, i, z, v);
         writeln('x', i: 1, ' = ', det(n, z) / den: 6: 6)
     end
end.
```

Figure 6.10

you get the printout for x_j as

1   8   1   −1   9   −9

respectively.

The method of *Gaussian elimination* is sometimes used in matrix reduction problems. The idea is that if a matrix can be triangularized by row-column reduction, the determinant is simply the product of the diagonal elements. The program "reduce" in Figure 6.11 illustrates this concept, as does the "dot" function of Appendix B.

The procedure "solve" in the matrix library described in Appendix B makes use of many of the ideas discussed in this chapter. The idea behind the procedure is that you specify the coefficients as an $n \times n$ matrix a and the constants as an n column c. Then the call

solve(n,a,c,x)

will force x to be a solution to the n equations.

```
program reduce(input, output);
(* this program solves m simultaneous equations in m unknowns of the    *)
(* form a[i,1]*x[1]+...+a[i,m]*x[m]=y[i]   where i=1..m and m<=15 *)
(* the method of gaussian elimination with partial pivoting is used *)

const
     r = 15;
     c = 16;
type
     matrix = array [1..r, 1..c] of real;
     vector = array [1..r] of real;
var
     a: matrix;
     x: vector;
     i, j, k, l, m, nxt: integer;
     cn, big, term, temp, piv, sum: real;

     procedure getmat(var a: matrix);
     var
        j, k: integer;
     begin
        for j := 1 to m do begin
             writeln('enter row number ', j: 2);
             for k := 1 to m + 1 do
                  read(a[j, k])
        end
     end; { getmat }

begin
     write(' enter the number of equations, m= ');
     read(m);
     writeln;
(* read the enhanced coefficient matrix row by row including the y[i]   *)
     writeln('enter the enhanced matrix coefficients row by row');
     writeln('include the y(i) as the (m+1) entry in the (i)th row');
     writeln('enter the (m+1) numbers separated by blanks');
     writeln;
     getmat(a);
(* perform pivoting and triangularization of coefficient matrix *)
     for i := 1 to m - 1 do begin
        big := 0;
        for k := i to m do begin
             term := abs(a[k, i]);
             if term - big > 0 then begin
                  big := term;
                  l := k
             end
        end;
(*   check if zero pivot  term  has been found *)
        if big = 0 then
             writeln('a zero pivot element, terminate program');
(* if i <> l then switch the i-th and l-th rows   *)
        if i - l <> 0 then
             for j := 1 to m + 1 do begin
                  temp := a[i, j];
                  a[i, j] := a[l, j];
                  a[l, j] := temp
             end;
(* now start the pivotal reduction to triangular form *)
        piv := a[i, i];
        nxt := i + 1;
        for j := nxt to m do begin
             cn := a[j, i] / piv;
             for k := i to m + 1 do
                  a[j, k] := a[j, k] - cn * a[i, k]
        end
     end;
```

Figure 6.11

```
(* display the triangularized matrix *)
    writeln;
    writeln('the gaussian triangularized matrix is');
    for j := 1 to m do begin
        for k := 1 to m + 1 do
            write(a[j, k]: 12: 7);
        writeln
    end;
    writeln;
(* backsolve the equations for the x(i) *)
    writeln('the values of x(i) are as follows');
    for i := m downto 1 do begin
        sum := 0;
        for j := i + 1 to m do
            sum := sum + a[i, j] * x[j];
        x[i] := (a[i, m + 1] - sum) / a[i, i]
    end;
    for k := 1 to m do
        writeln('x(', k: 2, ')= ', x[k]: 14: 9)
end.
```

___

Figure 6.11 (continued)

Exercises

1. Find the inverse of the 6 × 6 matrix on page 99 by using procedure "invert" in the matrix library described in Appendix B. Verify numerically that the original matrix times its inverse is the identity matrix.

2. Write a program that asks for the three semiaxes of an ellipsoid, that is, the triplet (a,b,c) for the region

$$\frac{x_1^2}{a^2} + \frac{x_2^2}{b^2} + \frac{x_3^2}{c^2} \le 1$$

and prints out the volume. Modify the program so that it accepts the coefficients a_{ij} for the candidate ellipsoid

$$\sum a_{ij} x_i x_j \le 1$$

where i, j take values 1, 2, 3. Using these a_{ij}:
    (a) Decide whether this is a bounded region.
    (b) If bounded, find the volume.

The key to this exercise is to find relevant conditions on the matrix of a_{ij} to do task a, and then use the determinant of task a to answer task b. *Note:*

$$\text{Volume} = \frac{4\pi}{3 \sqrt{\det(a)}}$$

3. Attempt to generate large classes of *unimodular matrices,* that is, matrices with determinant 1.

4. Write a program to find the eigenvalues of an input matrix.

5. Investigate the $N \times N$ *Toeplitz matrices*

$$
T_N(A,B) =
\begin{bmatrix}
A & B & 0 & 0 & 0 & \cdots \\
B & A & B & 0 & 0 & \cdots \\
0 & B & A & B & 0 & \cdots \\
0 & 0 & B & A & B & \cdots \\
\vdots & & & \ddots & & \\
& & & & \ddots & B \\
\vdots & & & & B & A
\end{bmatrix}
$$

in several ways as follows. First, write a function that returns the determinant det[$T_N(A,B)$]. Then verify numerically, for some choice of N and various real numbers x that the value of

$$\det\{T_N[2\cos(x), -1]\}$$

is oscillatory in x. What is the period?

Second, work out on paper a recurrence relationship involving the determinants for consecutive dimensions $N - 1$, N, and $N + 1$; A, B fixed. Then redefine the determinant function using this relationship, and verify that computations of det are now much faster.

Finally, for fixed A, B find an $N \times N$ rotation matrix R such that $R^{-1}TR$ is purely diagonal. A good test of such a search is to show that the product of the pure-diagonal elements is the original determinant.

Answers

1. The inverse matrix is

$$
\begin{array}{rrrrrr}
1 & 0 & 0 & -1 & 0 & 1 \\
-1 & 0 & 2 & 0.5 & 0.5 & -2 \\
0 & 1 & -2 & 0 & 0 & 1 \\
0 & -2 & -2 & -0.5 & 0.5 & 0 \\
-1 & 1 & 0 & 0 & 0 & 0 \\
1 & 0 & -1 & 0.5 & -0.5 & 0
\end{array}
$$

2. The volume of the first ellipsoid is 4/3 * pi * abc. A region is bounded if the inverse of matrix a exists.

3. For 2×2 unimodular matrices, you can investigate the theory of equations $ad - bc = 1$ in a text on number theory. For large unimodular matrices, you can compute random determinants. Note that the group of such matrices is closed, so if you find two, their product is a third one, and so on.

4. The eigenvalues L of a matrix M are the zeros of the determinant $\det(M - LI)$, where I is the identity matrix. Zeros can be found using methods similar to those in exercise 3, page 89.

5. It turns out that

$$\det\{T_N[2 \cos(x),-1]\}$$

is equal to $\sin[(N+1)x]/\sin(x)$. The problem of finding R is difficult. A special case is $A = 2$, $B = -1$ for which a good diagonalizer has $R_{ij} = \sqrt{2/N+1} \sin(\pi ij/(n+1))$.

RECURSIVE FUNCTIONS AND PROCEDURES

In Chapter 2 you learned how to generate some sequences by using recurrence relationships. Whenever you program a recurrence relationship in Pascal, consider the possibility of *recursive* functions and procedures. These are blocks in a Pascal program that refer to themselves.

For example, a simple *recursive function* calculates *combinatorial brackets*

$$\binom{n}{m} = \frac{n!}{m! * (n-m)!}$$

as follows:

```
function combo(n,m: integer): integer;
      begin
            if m = 1 then combo := n
            else combo := (n/m) * combo(n − 1, m − 1);
      end;
```

Note that "combo" calls itself again and again, and that the calling stops when "$m = 1$" is satisfied, which is the key to the termination of the recursive calling. Before reading on, make sure you understand why the mainline statement

```
writeln(combo(6,2));
```

will write out the number 15, given that

$$15 = \binom{6}{2} = \frac{6 * 5}{2 * 1}$$

We do not want to compute factorials to get the bracket

$$\binom{n}{m}$$

for the reasons stated in Chapter 2, in the section titled *Approximating Limits*. Instead, notice that the factorials reduce to

$$\binom{n}{m} = \frac{n(n-1)\ (n-2)\ \cdots\ (n-m+1)}{m(m-1)\ (m-2)\ \cdots\ \qquad 1}$$

and this reduced formula is what motivated the structure for the function "combo".

Examples of recursive functions, notably the gamma function whose various reflection formulas allow efficient recursive programming, are given in Appendix D. Examples of applications of recursive functions are Fibonacci numbers (discussed in Chapter 2) and the generation of prime numbers.

We are used to the fact that there is no easy algorithm for generating the nth prime. It is possible, however, to generate small primes with a simple recursive function:

```
function f(n,p): integer: integer;
    begin
        if sqr(n) > p then f := 1 else begin
            if p mod n = 0 then f := n else f := f(n + 2, p);
        end;
    end;
```

It turns out that the quantity (given p is odd)

f(3,p)

equals 1 if and only if p is a prime. This recursive function was used to generate Figure 6.12, which consists of dots printed *lexicographically* (as you read row by row) corresponding to the positions of all primes less than 1,000,000. Striations and other *pseudopatterns* are evident, as is a logarithmic decrease in dot density.

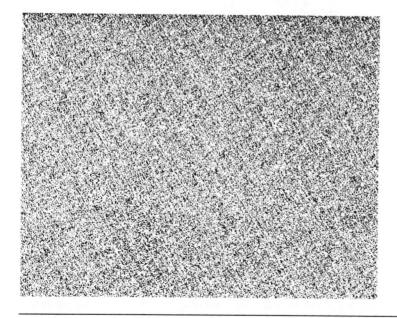

Figure 6.12 Lexicographic plot of all primes less than 1 million. Each prime is a dot. A gradual (logarithmic) decrease in dot density is evident.

```
program perm(input, output);
(* this program writes out all permutations on 1,...,size *)
const
    size = 10;
var
    n, i: integer;
    l: array [1..size] of integer;

    procedure switch(a, b: integer);
    var
        temp: integer;
    begin
        temp := l[a];
        l[a] := l[b];
        l[b] := temp
    end; { switch }

    procedure list;
    var
        k: integer;
    begin
        for k := 1 to size do
            write(l[k]: 1, ' ');
        writeln
    end; { list }

    procedure permute(k: integer);
(* this procedure is recursive *)
    var
        j: integer;
    begin
        if k = size then
            list
        else begin
            for j := k to size do begin
                switch(j, k);
                permute(k + 1)
            end
        end
    end; { permute }

begin
    for i := 1 to size do
        l[i] := i;
    permute(1)
end.
```

Figure 6.13

This density decrease is expected from the *prime number theorem*, which says that p is prime with approximate probability $(\log p)^{-1}$.

Recursive procedures can be used to rearrange Pascal entities. A classic example is the generation of all permutations on the set 1, 2, . . . , size, as shown in the program "perm" in Figure 6.13. The total number of permutations is size! (size factorial). This number grows very rapidly. In the program "perm", for example, the declared constant "size" on the order of 10 will cause extraordinarily long waits for output because of the magnitude of 10!. This program is similar to the prime generation program that printed Figure 6.12, in that you call the recursive block with a constant argument involved [in the program "perm", "perm(1)" is called in the main block].

Another example of recursive procedures is the drawing of recursively defined curves, such as the C curve obtained from the following algorithm:

1. Start with a given line segment.
2. If the line segment is shorter than a certain minimum, draw it.
3. If the line segment is longer than a certain minimum, draw two equal, perpendicular lines, one at each of the endpoints of the line segment in step 1.
4. Apply step 2 to the new lines.

Return to step 2. The program "recdraw" in Figure 6.14 generates C curves in this manner. Work through the steps executed by the program with a pencil to

```
program recdraw(input, output);
(* Recursively draw 'C' curve
   using user-supplied parameters. *)

const
    pi = 3.141592653;
    pidiv4 = 0.78539816095;
    sqrt2 = 1.41421356;
var
    x, y, length, angle, minlength: real;

#include "plibh.i"

    procedure linedraw(len, ang: real);
    (* polar draw *)
    begin
     x := x + len * cos(ang);
     y := y + len * sin(ang);
     draw(x, y)
    end; { linedraw }

    procedure ccurve(lngth, angl: real);
    (* recursively make curve *)
    begin
     if lngth <= minlength then
         linedraw(lngth, angl)
     else begin
         ccurve(lngth / sqrt2, angl + pidiv4);      (*first half*)
         ccurve(lngth / sqrt2, angl - pidiv4)       (*second half*)
     end
    end; { ccurve }
begin
    write('minimum line length: ');
    readln(minlength);
    write('length of beginning vector: ');
    readln(length);
    write('vector angle: ');
    readln(angle);
    write('initial coordinates: ');
    readln(x, y);
    graph;
    clear;
    move(x, y);
    ccurve(length, angle);
    alpha
end.
```

Figure 6.14

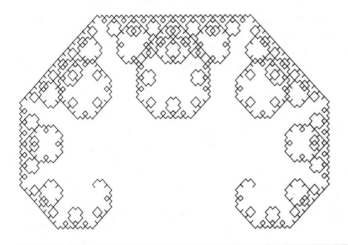

Figure 6.15   Recursively defined "C" curve, drawn by the program "recdraw" in Figure 6.14.

mimic the program's first few segment drawings. Figure 6.15 shows a typical C curve drawn with the program "recdraw".

PRECISION ARITHMETIC

In certain situations, we want more precision than normal Pascal implementations allow. For example, we may wish to analyze the convergence for a special function or series to 50 or more decimals or we may wish to work with extremely large integers, for example, 50 digits or more. The program "high" in Figure 6.16 is designed for high-precision arithmetic. It adds two long number strings in any base through base 10.

The key to this type of a program is the following loop:

```
repeat
     sum := x[ptr] + y[ptr] + car;
     car := sum div base;
     x[ptr] := sum mod base;
     ptr := ptr + 1;
until ptr := prec;
```

This loop illustrates the addition algorithm that we normally perform from right to left. Variables "car" is the *carry*, "ptr" is a *pointer* to the next position of the summand arrays, x, y, and so on. Multiplication, division, and subtraction are carried out in similar ways. In fact, multiplication can call "add" from time to time and division can call a "subtract" routine from time to time. Subtract itself can use "add" and a negation routine.

Notice that in the program "high" there is a user-defined type "num", so that in many respects the logic is similar to that for vectors (Chapter 3) and for complex numbers (section titled *Complex Numbers*).

```
program high(input, output);
(* this program asks for base and adds two input strings *)
const
    prec = 49;
type
    num = array [0..prec] of integer;
var
    x, y, z: num;
    base: integer;

    procedure list(x: num);
    var
        j: integer;
        lead: boolean;
    begin
        lead := false;
        for j := prec downto 0 do begin
            if x[j] <> 0 then
                lead := true;
            if lead then
                write(x[j]: 1)
        end;
        writeln
    end; { list }

    procedure getnum(var x: num);
    var
        ch: char;
        j, ctr: integer;
        temp: num;
    begin
        ctr := 0;
        while not eoln do begin
            read(ch);
            temp[ctr] := ord(ch) - ord('0');
            ctr := ctr + 1
        end;
        for j := 0 to ctr - 1 do begin
            x[j] := temp[ctr - 1 - j]
        end;
        readln
    end; { getnum }

    procedure add(var x: num; y: num);
    var
        car, sum, ptr: integer;
    begin
        ptr := 0;
        car := 0;
        repeat
            sum := x[ptr] + y[ptr] + car;
            car := sum div base;
            x[ptr] := sum mod base;
            ptr := ptr + 1
        until ptr = prec
    end; { add }
    begin
        write('base: ');
        readln(base);
        getnum(x);
        getnum(y);
        add(x, y);
        list(x)
    end.
```

Figure 6.16

Exercises

1. Write a general program to multiply, divide, and otherwise manipulate arbitrarily long integers. Extend the program to reals.
2. Write a program to compute and display exp(x) to 80 decimals. Calculate $\exp[\pi * \mathrm{sqrt}(163)]$ and notice what a beautiful number this is.
3. Write a program to calculate the continued fraction for a positive real x in the form

$$x = a_0 + \cfrac{1}{a_1 + \cfrac{1}{a_2 + \cdots}}$$

where the a_j's are positive integers. The set $(a_0, a_1, a_2, \ldots)$ is unique for the given x. Verify the celebrated result that every number of the form $A + \mathrm{sqrt}(B)$, for A, B rational eventually gives a *periodic sequence* of a_j. You may also be interested in the sequence for $x = (1 + e)/(1 - e)$.

Answers

1. This exercise is exploratory.
2. This exercise is exploratory. Extension to reals can be done in several ways. One way is to mimic floating-point processing, in which a mantissa and exponent are kept separate throughout almost all of the computation, which is equivalent to computation in scientific notation.
3. This exercise is exploratory.

ORTHOGONAL POLYNOMIALS

Useful orthogonal polynomials and applications follow:

Legendre polynomials: Function "leg(n: integer; x: real): real". These are useful for expansions in spherical coordinates and for certain differential equations (Chapter 2 references: Abramowitz and Stegun, 1965). Standard: $P_n(x)$.

Tchebyshev polynomials: Function "tcheb(n: integer; x: real): real" These are useful for expansion of continuous functions over finite intervals. Standard: $T_n(x)$.

Laguerre polynomials: Function "lag(n: integer; a,x: real): real". These are useful for expansion of continuous functions and for certain differential equations of physics and chemistry. Standard: $L_n^{(a)}(x)$.

Hermite polynomials: Function "her(n: integer; x: real): real". These are useful for expansion of continuous functions on $(-\infty, \infty)$ and for certain differential equations, especially oscillator equations. Standard: $H_n(x)$.

Jacobi polynomials: Function jac(n: integer; a,b,c: real): real". These are useful in ultraspherical geometry problems and for certain differential equations. Standard: $P_n^{(a,b)}(x)$.

Exercises

For these exercises, either use functions from Appendix D files or regenerate the functions yourself.

1. Write a program that evaluates the integrals

    $$\int_R W(x)Q_m(x)Q_n(x)\ dx$$

    for the various orthogonal polynomials Q_m, Q_n (of the same class) with their associated weight functions W and regions for integration R. For example, Legendre: $R = (-1,1)$, $W(x) = 1$; Hermite: $R = (-\infty, \infty)$, $W(x) = \exp(-x^2)$; Jacobi: $R = (-1,1)$, $W(x) = (1 - x)^a(1 + x)^b$; Laguerre: $R = (0,\infty)$, $W(x) = x^a\exp(-x)$; Tchebyshev: $R = (-1,1)$, $W(x) = (1 - x^2)^{-1/2}$. In all of these cases if $m \neq n$, you should obtain zero, otherwise you should obtain an expression you can check in Chapter 2 reference: Abramowitz and Stegun (1965). These observations are useful in approximating curves with the polynomials, as in exercises 2, 3, and 4 below.
2. Write a program to expand an arbitrary function $f(x)$ on the real line into a sum of terms $c_n H_n \exp(-x^2/2)$, where the task is to choose the c_n. *Hint:* Use the result in exercise 1 and consider

    $$\int_{-\infty}^{\infty} f(x)H_m(x)\ \exp\left(\frac{-x^2}{2}\right)\ dx$$

    for arbitrary m. This hint applies for all orthogonal polynomial expansions.
3. Write a program to expand arbitrary polynomials as linear combinations of Tchebyshev polynomials over the interval $(-1,1)$.
4. Write a program to expand arbitrary functions $f(r,\theta,\phi)$ (spherical coordinates) using Laguerre polynomials, Legendre polynomials, and weight functions.
5. Work the above ideas into a general *spline curve* program. The program should take a set of arbitrary data points and draw a smooth curve through all of them. Legendre polynomials are good for generating splines on the interval $(-1,1)$.

Answers

1. This exercise is exploratory.
2. The integral is in fact equal to the desired coefficient c_m.
3. Use the integral method of exercise 2 to get coefficients.

4. This exercise is exploratory. One function to try is

   f(r,theta,phi) = exp(−sqr(r)) * cos(theta) *exp(−theta)

5. If you consider f(x) as a sum of Legendre polynomials $P_n(x)$ times respective coefficients a[n] and use a total number of terms equal to the number of data points (x,y), you have N equations in N unknowns a[N] and can solve for the coefficients using matrix methods in the section titled *Linear Equations*.

SPECIAL FUNCTIONS

Special functions available in Appendix D and brief applications are listed here:

Bessel function: Function "j(nu,x: real): real". These satisfy Bessel's differential equation, are useful in studies of asymptotics, and generate other functions, such as Airy functions. Standard: $J_{nu}(x)$.

Gamma function: Function "gam(x: real): real". The gamma function appears in the definition of many other functions and is a superb analytic function for testing programs that involve complex numbers. Standard: $\Gamma(x)$.

Complex error function: Function "erfr(x,y: real): real" and function "erfi(x,y: real): real". This function is given in two parts, real and imaginary, respectively. Uses include statistical calculations and solutions to certain differential equations (related to Fresnel integrals). Standard: erf(x + iy).

Modified Bessel function: Function "i(nu,x: real): real". This is related to the standard Bessel function $J_{nu}(x)$ and is in a certain sense the analytic continuation of J_{nu}. Standard: $I_{nu}(x)$.

Modified Bessel function: Function "k(nu,x: real): real". This is related to the standard Bessel function $Y_{nu}(x)$. Standard: $K_{nu}(x)$.

Gauss hypergeometric function: Function "f(a,b,c,z: real): real". This is used to create many of the above functions.

Kummer hypergeometric function: Function "m(a,b,z: real): real". This function is used for creating other functions.

Associated confluent hypergeometric function: Function "u(a,b,z: real): real". This function is used for creating other functions.

These special functions have many applications, such as the ones listed below:

1. Practice with graphics. Figure 6.17 shows a graph of the imaginary part of the complex error function, which we call "erfi(x,y)".
2. Expansion of functions in series of special functions.
3. Analysis of more functions, for example, the *Riemann zeta function,* which is not in the Appendix D library but whose proper study involves gamma and other functions.
4. Solutions of differential equations. Bessel functions, for example, are known to solve many types of differential equations. The so-called WKB-type approximations to differential equations involve the *Airy function,* which is related to the Bessel function of order nu = 1/3.

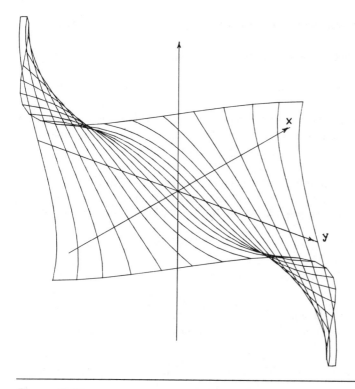

Figure 6.17

Exercises

1. Attempt to work out *hidden lines* graphics routines, which we do not cover in this book. Start with the graph of "erfi" (Figure 6.17), which has all the properties to make it a good test case.

2. Extend one or more of the special library functions in Appendix D to the complex plane. Some functions can then be dropped because of interrelationships on the complex plane that cannot be exploited in real calculations. An example is the relationship of Bessel functions to modified Bessel functions.

3. Study the zeros of Bessel functions, with or without graphics. For example, j(n,x) for integer n has an infinite set of zeros that, for large x, become equally spaced.

4. Solve the differential equation

$$\frac{d^2x}{dt^2} = xt$$

and compare it with calculations of the Airy function

$$Ai(t) = \frac{t \exp(-2t^{3/2}/3)}{3^{5/6} \sqrt{\pi}} \, 2^{2/3} \, u\left(\frac{5}{6}, \frac{5}{3}, \frac{4t^{3/2}}{3}\right)$$

where u is the associated hypergeometric function in the Appendix D library.

5. The complete elliptic integral of the first kind is

$$K(m) = \int_0^{\pi/2} (1 - m^2 \sin^2 t)^{-1/2} \, dt$$

and is known to be equal to $(\pi/2)f(1/2,1/2,1,m)$, where f is the Gaussian hypergeometric function of Appendix D. Work out a way to do the integral as accurately as f gives it. See exercise 3, page 39 since K there is in fact $K(1/2) * 2/\pi$.

6. It is known that you cannot get the exact perimeter of an arbitrary ellipse. For certain eccentricities, however, it is possible to obtain closed expressions. Perform numerical integration for the perimeter P of the ellipse given in polar coordinates by

$$r = \frac{1 - E^2}{1 - E \cos \theta}$$

where E is the constant $\sin(\pi/12)$. Using the gamma function in Appendix D, compare to the exact result of Ramanujan (Whittaker and Watson, 1972):

$$P = \frac{\pi^{1/2}}{3^{1/4}} \left((1 + 3^{-1/2}) \frac{\Gamma(1/3)}{\Gamma(5/6)} + 2 \frac{\Gamma(5/6)}{\Gamma(1/3)} \right)$$

7. Work out the exact expression for the volume of the N-dimensional unit ball in terms of gamma functions. By computer, find the (nonintegral) dimension N for which the volume is a *maximum*. (See exercise 5, page 57.)

Answers

1. This exercise is exploratory. One way to do hidden lines is to track along a line, starting from an arbitrary point on the surface, going straight to the viewer's eye, and deciding not to draw the surface point if more surface is encountered along the ray track.

2. Chapter 2 reference: Abramowitz and Stegun (1965) contains many of the complex-plane relationships, such as the Bessel relationship.

3. The zeros are listed in Chapter 2 reference: Abramowitz and Stegun (1965) for j(nu,x). The zeros become arithmetically spaced for large arguments x.

4. The Airy function should exhibit oscillatory behavior for large negative t and exponentially damped behavior for large positive t.

5. The number $K(1/2) * 2/pi$ is about 1.18, which is a good check on your programs. The *arithmetic-geometric mean* (AGM) approach for the elliptic integrals is interesting to program. See Chapter 2 reference: Abramowitz and Stegun (1965).

6. The arc length is 6.17660198 . . . .

7. The dimension having maximum volume is approximately 5, but is not an integer.

NUMBER THEORY

Two important kinds of operations for number theory calculation are as follows:

gcd(a,b: integer): integer;   This is the *greatest common divisor* (gcd) of the two numbers a, b. We interpret this function to be unchanged by a sign change of a or b or both. For convenience, we return zero if a $= 0$.

pmod(x,y,z: integer): integer;   This is $x^y (\mod z)$ and is computed as quickly as is possible in a simple Pascal block.

Both tests are useful in investigations of the following:

1. Prime numbers, which can be tested for primality with the mod function and sometimes with the gcd.
2. Numbers to be factored; we shall see examples of how the gcd is used in this situation.
3. Determination of efficient decimation for fast Fourier transforms or auto-correlations.
4. Pseudorandom sequences and noise.

The function "pmod" allows arithmetic involving much larger primes than were generated in the section titled *Recursive Functions and Procedures*. The routine in Figure 6.18 is almost instantaneous to the user, even for primes with tens of digits, because the test performed takes on the order of $(\log p)^3$ steps (Crandall and Penk, 1979; Buhler et al., 1982). This routine will return, for example, "pmod(2,102,103)" as 1. In fact, 103 is prime and the *Fermat test*, that of checking "pmod(2, p − 1, p)" for prime candidates p, is passed (result $= 1$). When the result is not 1, p is definitely a *composite* number (nonprime); when the result is 1, p is probably prime (but may not be).

The program "primetest" in Figure 6.19 tests candidates in this way. The input numbers p can be arbitrarily large, limited only by the precision of your

```
    function pmod(x, y, z: integer): integer;
(* return x ^ y (mod z) *)
    var
       e: integer;
    begin
       e := 1;
       while y <> 0 do begin
          if y mod 2 <> 0 then begin
             y := y - 1;
             e := e * x mod z
          end;
          y := y div 2;
          x := x * x mod z
       end;
       pmod := e
    end; { pmod }
```

Figure 6.18

```
program primetest(input, output);
(* when you input a number p, it is composite if output <> 1 *)
(* if output is 1, p is very likely but not necessarily prime *)
var
    p: integer;
#include "num.i"
begin
    repeat
        readln(p);
        writeln(pmod(2, p - 1, p))
    until false
end.
```

Figure 6.19

machine. The "include" file "num.i" contains the block "pmod(2, p − 1, p)" that defines "pmod". When this program is run, and you type in

11111

the answer is 10536, meaning "11111" is composite.

The "gcd" function is implemented as shown in Figure 6.20. The algorithm is Euclid's algorithm, and is, like "mod", rather fast in the sense that the gcd of two numbers of tens of digits each is computed in a few seconds. A routine that makes practical use of the gcd is the *factoring routine* "pollard" in Figure 6.21. It is a

```
    function gcd(a, b: integer): integer;
(* return gcd of a and b *)
    var
        q, r: integer;
    begin
        if b < 0 then
            b := -b;
        if a < 0 then
            a := -a;
        if a > 0 then begin
            b := b mod a;
            r := 1;
            if b = 0 then
                r := 0;
            while r > 0 do begin
                q := a div b;
                r := a - q * b;
                a := b;
                b := r
            end
        end;
        gcd := a
    end; { gcd }
```

Figure 6.20

```
program pollard(input, output);
(* input a number p to be factored using 'pollard rho' method *)
var
    x, y, p: integer;
#include "num.i"
begin
    repeat
        readln(p);
        x := 3;
        y := 3;
        while gcd(y - x, p) < 2 do begin
            write('.');
            x := (x * x + 2) mod p;
            y := (y * y + 2) mod p;
            y := (y * y + 2) mod p
        end;
        writeln(gcd(y - x, p): 1)
    until false
end.
```

Figure 6.21

---

modern method, based on the *Pollard rho algorithm.* Again, "num.i" contains the "gcd" function. When this program is run and you input

167539

the result is first a series of dots and then a factor of the input number, namely 701.

Care must be taken in such programs not to exceed maximum integer size on your machine. For large integer calculations, you can implement arbitrary-precision multiply and divide, using ideas in the section titled *Precision Arithmetic.*

Exercises

1. Write a program to find composite numbers p having pmod(2, p − 1, p) = 1. Such a number is a base 2 *pseudoprime* of which there are fewer than 1000. A true pseudoprime has pmod(a, p − 1, p) = 1 for each a = 1, ..., p − 1; this is sometimes called a *Carmichael number.* Find a Carmichael number.
2. Write a program that takes an arbitrary positive integer as input and writes out all factors, for example,

    525 (return)
    3*5*5*7

    Try to factor 491401.
3. Write a program that does *not* use the "sqrt" function but does determine whether an input integer is a *perfect square.* Write modifications to decompose an arbitrary integer into the sum of four squares, for example,

$66 = 0^2 + 1^2 + 4^2 + 7^2$. This is always possible according to a famous theorem of Legendre. Are there Fibonacci numbers that are perfect squares?

4. Verify that the sum of the reciprocals of the primes less than n grows like $\ln[\ln(n)]$.

5. Write a factoring program based on the principle that if p is to be factored and $p + n^2$ is a perfect square m^2, then $p = (m + n)(m - n)$.

6. A *triangular number* is one in the sequence 1, 3, 6, 10, . . . , or, in general, $n(n + 1)/2$. Is every number (within some large range) expressible as the sum of a triangular number and a prime number?

7. Write a program to find the *least common multiple* (lcm) of two integers. For example, lcm(10,6) = 30. Then attempt to verify that the lcm of all the integers 1, . . . , n grows exponentially in n, where the lcm of more than two integers is as follows: the lcm of a, b, c, etc., is lcm[lcm(a,b),c], etc., recursively.

Answers

1. The number 541 is a Carmichael number.

2. The factors of 491401 are 701*701.

3. There are many ways to determine if an integer is a perfect square. One way is to check whether every prime p dividing a number does so with a highest power that is *even*. For example, 1875 would not be a square because 5's highest power 3 is an odd power. The number 144 is a square Fibonacci number.

4. This is a standard number theory result.

5. You can use the methods in exercise 3 or the sqrt function as follows:

   if sqr(trunc(sqrt(n))) = n then square := true;

   to test whether n is a square.

6. There is one number, apparently, that cannot be expressed as the sum of a triangular and prime number. It is greater than 200 and less than 300. No results have been rigorously proven, however.

7. This exercise is exploratory. The result is exp(n) for asymptotically large n.

7 | Examples from Chemistry

Now there is one outstandingly important fact regarding Spaceship Earth, and that is that no instruction manual came with it.

R. Buckminster Fuller
Operating Manual for Spaceship Earth (1970)

STOICHIOMETRY

There are many books that tell you how to write Pascal programs, including this book, but no one will tell you *when* to write them.

Computer languages are similar to Spaceship Earth in that they do not come equipped with applications criteria. To illustrate this dilemma, consider an example from chemistry.

Stoichiometry problems require us to balance chemical reactions (Bailar et al., 1978) such as the following:

$$A \ H_2 + B \ O_2 \rightarrow C \ H_2O$$

It is easy to see that for this reaction to be balanced, $A = C$ and $B = A/2$. Do you have to write a program to work this out? Probably not; even for complicated products you can easily work the balance in your head or on paper. A program would be more trouble than it is worth, unless of course there is a catch.

Suppose that information about the products is not complete, in which case you have to find which of several balanced equations yields the proper laboratory data. Assume that an *unknown* gaseous hydrocarbon is burned and you know the following:

1. The density of the original gas sample, in grams per liter
2. The mass of the H_2O product, in grams
3. The mass of the CO_2 product, in grams

You want to guess the formula of the original hydrocarbon. For this type of chemistry problem it is helpful to use a program in Pascal, such as the one in Figure 7.1. The program would be most useful if many such reactions had to be

```
program combustion(input, output);
(* program by R. Whitnell.       *)
(* combustion analysis of hydrocarbons - uses amount of
    h2o, co2 produced when sample is burnt, to deduce
    chemical formula of original compound *)
const
    caw = 12.0;
    haw = 1.0;
    co2mw = 44.0;
    h2omw = 18.0;
    ideal = 22.4;                  (* volume of one mole of ideal gas *)
var
    co2, h2o, hyd, car, rho, ratio, hatoms, wtgas, catoms: real;
begin
    write('density of sample to be burnt, gms/liter?  ');
    readln(rho);
    write('grams of h20 produced?  ');
    read(h2o);
    write('grams of co2 produced?  ');
    read(co2);
    hyd := 2 * h2o / h2omw;
    car := co2 / co2mw;
    ratio := car / hyd;
    wtgas := rho * ideal;
    catoms := wtgas / (haw * ratio + caw);
    hatoms := catoms / ratio;
    writeln('best guess formula is: ');
    writeln('c:', catoms: 1, '  h:', hatoms: 1)
end.
```

Figure 7.1

methodically analyzed or if the reactions were much more complex than our example.

When the program "combustion" in Figure 7.1 is run, you can enter responses to the program statements as follows:

density of sample to be burnt, gm/liter?   1.2
grams of h_2o produced?   2.0
grams of co_2 produced?   3.5

The output in this case is

best guess formula is:
c = 2.18   h = 6.08

and the gas is best guessed to be C_2H_6.

The rationale behind the program is as follows. Assuming the initial gas is ideal, the molecular weight (MW) is

$$MW = (rho)*22.4 \text{ liters/mole}$$

where rho is the density of the gas sample. Now the molecular weight is

$$MW = 12C + H$$

where C and H are the respective numbers of atoms in the unknown molecule. If we define

$$ratio = C/H$$

which is also the ratio of moles CO_2 to moles H_2O, it follows that

$$C = MW/(ratio + 12)$$
$$H = C/ratio$$

which are the output variables of the program.

REACTION MODELS

Applications of probabilistic models abound in biology, chemistry, and physics. Often the application involves thermodynamical considerations, especially when *statistical ensembles* of atoms, molecules, or the like are being modeled. The random walk and Monte Carlo techniques discussed in Chapter 4 are often helpful in these modeling problems. Consider the simple reaction

$$A \rightarrow B + C$$

Theory of reactions, which is based on thermodynamics, tells us that the *concentrations* should obey the limiting differential equations.

$$\frac{d[A]}{dt} = -k_a[A]; \text{ in absence of B and C}$$

$$\frac{d[B]}{dt} = \frac{d[C]}{dt} = -k_{bc}[B][C]; \text{ in absence of A}$$

where k_a and k_{bc} are *rate constants*. These *time derivatives* take the forms shown here when reaction in the opposite direction is not significant, hence the phrase "in absence of . . . ".

Let us model this simple system *microscopically* in the sense that numbers of molecules will be small. In such a situation time derivatives have no meaning. Time derivatives work in the standard theory because molecular counts are typically enormous.

Consider the following operations that are executed on each pass of some loop:

1. Have *one molecule* of A react (to form B and C) with probability $+k_a[A]$
2. Have *one molecule* each of B and C react (to form A) with probability $+k_{bc}[B][C]$

For programming simplicity, we will not establish any difference between concentration [A] and molecular count A in this problem. It is important to *scale* such computer calculations appropriately. It is left to you to scale results for the program "mix" in Figure 7.2, which performs the single-molecule looping given above. You have to ask typical scaling questions such as:

1. What is the *real time* corresponding to one pass of the loop?
2. What is the relationship between the k_a, k_{bc} of the program and the standard molar rate constants in standard units?

Do you see how questions 1 and 2 are closely related? The reason is that the loop operations refer to probability per unit time. Work out these scaling concepts if you wish to use such modeling in the future.

Notice the procedure "react" in the program "mix" in Figure 7.2. It is structured so that expansion to greater numbers of reactants is possible. The main modification for more reactants is the statement

```
type compound = (a,b,c,d, ... )
```

Small modifications must also be made in the procedure "react" and other parts of the program.

The "for-do" loops in the procedure are particularly easy when you use *scalar types,* for example, if you end the compound set with reactant "g" you can write statements such as

```
for cc := a to g do begin
```

The program "mix" has explicit calls to the "plibh.i" library package (in Appendix A. This allows graphics output on a Hewlett-Packard 7470A plotter. A typical output specimen is shown in figure 7.3. The starting data are

Number of A molecules = 20
Number of B molecules = 35
Number of C molecules = 16

You can see the tendency toward exponential growth and decay in reaching equilibrium. *Equilibrium* has the probabilities of the loop equal, so that you recover the familiar rate relationship:

$$\frac{[B][C]}{[A]} = \frac{k_a}{k_{bc}} = \frac{0.02}{0.008} = 2.5$$

Indeed, the *steady state,* which is somewhat rough in the graphics output because of the microscopic nature of the model, has $[B] \approx 23$, $[C] \approx 3$, and $[A] \approx 33$, which gives a reasonable ratio of 2:1.

SOLUBILITY CALCULATIONS

Solubility problems are good examples of how Pascal programs can be used for computational efficiency. An ionic solid M_mX_n goes into solution according to the equation

```
program mix(input, output);
(* microscopic model of the reaction a<>b+c *)
(* user inputs rates ka, kbc and initial numbers of a,b,c molecules *)
type
    compound = (a, b, c);
    mix = array [compound] of integer;
var
    lastsoup, soup, initsoup: mix;
    cc: compound;
    x, y, ka, kbc: real;
    t: integer;
#include "plibh.i"

    procedure react(var soup: mix; inc: integer);
(* increment reactant b,c counts by inc, and decrement #a by inc *)
    begin
        soup[a] := soup[a] - inc;
        soup[b] := soup[b] + inc;
        soup[c] := soup[c] + inc
    end; { react }

begin
    write('ka   kbc : ');
    write('#a   #b   #c : ');
    readln(ka, kbc);
    for cc := a to c do begin
        read(soup[cc]);
        initsoup[cc] := soup[cc];
        lastsoup[cc] := soup[cc]
    end;
    (* copy of initial numbers now saved *)
    graph;
    move(-1, 1);
    draw(-1, -1);
    draw(1, -1);
    move(1.1, -1);
    alphal;
    writeln('time');
    move(-0.8, 0.96);
    alphal;
    writeln('MICROSCOPIC REACTION MODEL');
    move(-0.5, 0.9);
    alphal;
    writeln('ka = ', ka: 5: 5, '; kbc = ', kbc: 5: 5);
    move(-1.18, 0);
    alphal;
    writeln('A=', soup[a]: 1);
    move(-1.18, -1 + soup[b] / soup[a]);
    alphal;
    writeln('B=', soup[b]: 1);
    move(-1.18, -1 + soup[c] / soup[a]);
    alphal;
    writeln('C=', soup[c]: 1);
    t := 0;
    repeat
        t := t + 1;
        if random(1) < soup[a] * ka then
            react(soup, +1);
        if random(1) < soup[b] * soup[c] * kbc then
            react(soup, -1);
        (* now plot the reactant counts *)
        for cc := a to c do begin
            x := -1 + (t - 1) / 100;
            y := lastsoup[cc] / initsoup[a] - 1;
            move(x, y);
            y := soup[cc] / initsoup[a] - 1;
            draw(x, y);
            x := -1 + t / 100;
            draw(x, y);
            lastsoup[cc] := soup[cc]
        end
    until t > 200
end.
```

Figure 7.2

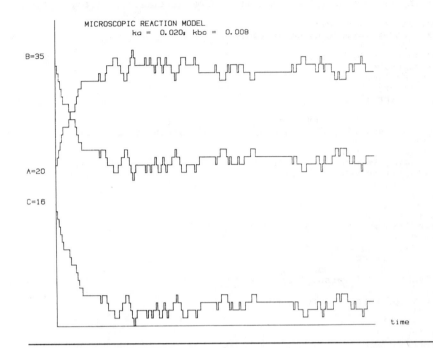

Figure 7.3

$$M_m X_n \rightarrow m M^{n+} + n X^{m-}$$

so that there are m cations and n anions in solution. The equilibrium constant for the reaction is

$$k_{sp} = [M]^m [X]^n = [M]^{m+n} (m/n)^n$$

For every mole of solid that dissolves, m moles of M ions appear in solution so that solubility of the solid $M_m X_n$ is [M]/m. Therefore,

$$\text{Solubility} = \frac{[M]}{m} = m^{-1} k_{sp}^{1/(m+n)} \left(\frac{m}{n}\right)^{n/(m+n)}$$

Calculations based on this expression are best done with a power procedure as described in Chapter 2 in the section titled *Functions*. The program "solubility" in Figure 7.4 carries out the required exponentiation steps. Typical input for the program is as follows:

for calcium fluoride:
1 cation
2 anions
ksp = 1.7e − 10

The solubility calculated should be 3.4e − 4.

```
program solubility(input, output);
(* compute in a straightforward manner the solubility of a substance
   MmXn having thus m cations and n anions *)
(* program by R. Whitnell *)
var
    ksp, sol: real;
    cat, an: integer;

    function pow(a, b: real): real;
    begin
    pow := exp(b * ln(a))
    end; { pow }

begin
    write('how many cations are present in compound?  ');
    read(cat);
    write('how many anions are present in compound?  ');
    read(an);
    write('what is solubility product, ksp ?  ');
    read(ksp);
    sol := pow(cat / an, an / (cat + an)) * pow(ksp, 1 / (cat + an));
    sol := sol / cat;
    writeln('solubility is ', sol)
end.
```

Figure 7.4

Exercise

Extend the program "solubility" in Figure 7.4 to handle solubility calculations when more species are present.

Answer

This exercise is exploratory.

pH CALCULATIONS

Let HA be a weak acid, K_a be the equilibrium constant for the reaction

$$HA + H_2O \rightarrow H_3O^+ + A^-$$

and $[HA]_0$ be the initial concentration of the acid, that is, before dissociation has occurred. If we denote the concentration $[H_3O^+]$ by X, we can write (Mahan, 1969) the equation

$$\frac{X(X - K_w/X)}{([HA]_0 - X + K_w/X)} = K_a$$

where $K_w = 1.0e - 14$ is the equilibrium constant for

$$2\,H_2O \rightleftharpoons H_3O^+ + OH^-$$

```
program ph(input, output);
(* solve for concentration [H3O+] by Newton's method *)
(* program by R. Whitnell *)
const
    kw = 1e-14;
var
    ka, hao, next, last: real;

    function pow(a, b: real): real;
    begin
    pow := exp(b * ln(a))
    end; { pow }

begin
    write('initial acid concentration (in molar)?  ');
    read(hao);
    write('pka of acid?  ');
    read(ka);
    ka := pow(10, -ka);
    write('enter initial guess at [h3o+]  ');
    read(last);
    repeat
    next := last - (pow(last, 3) + ka * sqr(last)
                        - last * (ka * hao + kw) - ka * kw)
                    / (3 * sqr(last) + 2 * ka * last - ka * hao - kw);
    last := next
    until abs(pow(next, 3) + ka * sqr(next)
            - next * (ka * hao + kw) - ka * kw) < 1e-15;
    writeln('[h3o+] = ', next);
    writeln('ph = ', -(ln(next) / 2.303): 7: 3)
end.
```

Figure 7.5

This example is ideal for *Newton's method,* discussed in Chapter 3, because X
satisfies a *cubic equation.* The program "ph" in Figure 7.5 solves the cubic
equation.

When the program "ph" is run, you can find the pH for *acetic acid* by entering
responses to the program statements as follows:

initial acid concentration (in molar)? 0.01
pka of acid? 4.733
enter initial guess at [h3o+] 1.0e − 3

The answer comes out as pH = 3.375.

Exercises

1. Use the technique of the program "ph" in Figure 7.5 to graph many values for
   acetic acid; let axes denote pH and initial acid concentration.
2. Extend the method used in exercise 1 to handle *diprotic* and *triprotic*
   acids.

Answers

1. This exercise is exploratory.
2. This exercise is exploratory.

TITRATION

A program that draws *titration curves* is a good example of a situation in which user input substantially determines the graphics output. For titration, *weak* and *strong acids* are handled differently; the program "titrate" in Figure 7.6 has those operations common to both types of acids.

For either a weak or strong acid, we define $\dot{F}$ as follows:

$$F = \frac{\text{number of moles of base added}}{n_0}$$

where n_0 is the initial number of moles of acid. For a strong acid and $f < 1$, it can be shown that

$$[H_3O^+] = \frac{n_0(1 - f)}{V + v} + \frac{K_w}{[H_3O^+]}$$

where V is the initial volume of the acid solution and v is the volume of base added. At $f = 1$ the solution is neutral. If more base is added so that $f > 1$,

$$[H_3O^+] = \frac{K_w}{[B]_0 v'/V'}$$

where v' is the volume added after $f = 1$ and V' is the total volume. For a weak acid with given pKa and $0 < f < 1$, it can be shown that

$$[H_3O^+] = \frac{(1 - f)K_a}{f}$$

whereas for $f > 1$ the weak acid follows the $[H_3O^-]$ relationship given above.

Two runs were done with the program "titrate" in Figure 7.6 to show typical output. The data are given in Table 7.1. Figure 7.7 shows the weak acid curve, and Figure 7.8 shows the strong acid curve.

Note the conditions in the "sketch" procedure of "titrate" that make sure the curves begin correctly. In graphics output programming, the *overhead,* that is, the conditions, questions, and so on, that prepare for picture drawing, is often much more work than the programming of the scientific operations.

Table 7.1 Data for the Program "Titrate"

Data requested	Weak acid	Strong acid
Molarity of solution to be titrated	0.1	0.1
Volume of solution to be titrated	50 ml	50 ml
Molarity of titrating base	0.1	0.1
Amount of base added each time	1 ml	1 ml
pKa	4.73	

```
program titrate(input, output);
(* plot titration curves :: program by R. Whitnell *)
const
    kw = 1e-14;
var
    pka, ka, moli, vol, voli, moltit, dvol, f, vole: real;
    no, b, xmax, xmin, ymax, ymin, hplus: real;
    ch: integer;

#include "plibh.i"
    function pow(a, b: real): real;
    begin
    pow := exp(b * ln(a))
    end; { pow }

    function ph(a: real): real;
    begin
    ph := -(ln(a) / 2.303)
    end; { ph }

    procedure drawaxes(xmin, xmax, ymin, ymax: real);
    var
    j: integer;
    temp: real;
    begin
    move(-1, -1);
    draw(-1, 1);
    move(-1, -1);
    draw(1, -1);
    for j := 1 to 5 do begin
        temp := -1.00 + (j - 1) * 0.5;
        move(temp, -1.00);
        draw(temp, -0.95);
        alphal;
        writeln(xmin + (j - 1) * (xmax - xmin) / 4: 5: 2)
    end;
    for j := 1 to 5 do begin
        temp := -1.00 + (j - 1) * 0.5;
        move(-1.05, temp);
        draw(-0.95, temp);
        move(-1.20, temp);
        alphal;
        writeln(ymin + (j - 1) * (ymax - ymin) / 4: 5: 2)
    end;
    move(-1.3, 0.2);
    alphal;
    writeln('pH scale');
    move(-0.1, -0.85);
    alphal;
    writeln('vol scale')
    end; { drawaxes }

    procedure sketch(x, y, xmax, ymax: real);
    begin
    if (y > ymax) or (abs(f) < 1e-5) then
        move(2 * x / xmax - 1, 2 * y / ymax - 1)
    else
        draw(2 * x / xmax - 1, 2 * y / ymax - 1)
    end; { sketch }
```

Figure 7.6

```
begin
    xmin := 0;
    xmax := 1.5;
    ymin := 0;
    ymax := 14;
    write('strong(1) or weak(2) acid?  ');
    read(ch);
    if ch = 2 then begin
    write('pka of acid?  ');
    read(pka);
    ka := pow(10, -pka)
    end;
    write('molarity of solution to be titrated?  ');
    read(moli);
    write('volume of solution to be titrated (in ml)?  ');
    read(voli);
    write('molarity of titrating base?  ');
    read(moltit);
    write('amount of base added each time (in ml)?  ');
    read(dvol);
    no := moli * voli;
    f := 0;
    vol := voli - dvol;
    graph;
    drawaxes(xmin, xmax, ymin, ymax);
    if ch = 1 then begin
    {titration curve for strong acid}
    repeat
        vol := vol + dvol;
        f := (vol - voli) * moltit / no;
        b := no * (1 - f) / vol;
        hplus := (b + sqrt(sqr(b) - 4 * kw)) / 2;
        sketch(f, ph(hplus), xmax, ymax)
    until f >= 0.98;
    vole := vol;
    repeat
        vol := vol + dvol;
        f := (vol - voli) * moltit / no;
        hplus := vol * kw / (moltit * (vol - vole));
        sketch(f, ph(hplus), xmax, ymax)
    until f >= 1.5
    end else begin
    {titration curve for weak acid}
    repeat
        vol := vol + dvol;
        f := (vol - voli) * moltit / no;
        if f < 0.999 then begin
            if abs(f) > 1e-5 then
                hplus := (1 - f) / f * ka
            else
                hplus := 1e14;
            sketch(f, ph(hplus), xmax, ymax)
        end
    until f >= 1;
    vole := vol;
    repeat
        vol := vol + dvol;
        f := (vol - voli) * moltit / no;
        hplus := vol * kw / (moltit * (vol - vole));
        sketch(f, ph(hplus), xmax, ymax)
    until f >= 1.5
    end
end.
```

Figure 7.6 (continued)

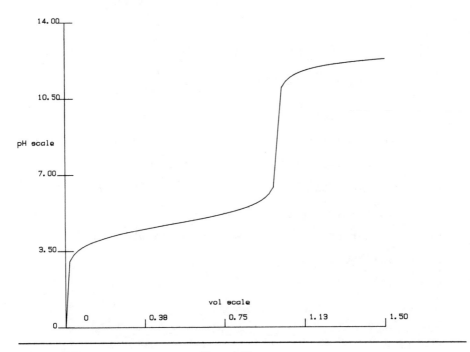

Figure 7.7 Weak acid titration, pKa = 4.73

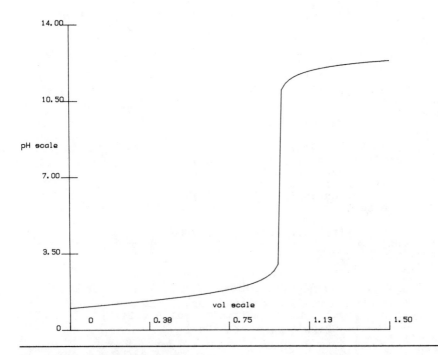

Figure 7.8 Strong acid titration.

Exercise

Titration curves of amino acids, which can be considered to some order diprotic, are of biochemical importance. Therefore, attempt to extend the problem of drawing tiration cruves to diprotic acids.

Answer

This exercise is exploratory.

STEREO MOLECULES

Graphics can be used to display models of stereo molecules. If perspective is used, the differences among various optical isomers can be readily seen. For these problems, apply what you learned about graphics in Chapter 5 and use library procedures from Appendix A.

The idea of *perspective* is that objects close to an observer appear to be magnified because of their proximity alone. Suppose that we want to view a point (x,y,z) at an orientation defined by Euler angles (a,b,c) (see Chapter 5) and that we also want perspective. One way to achieve this is as follows:

```
rotate(x,y,z,a,b,c);     (*this gets the proper view*)
x := warp*x;             (*these get new x,y for direct plotting*)
y := warp*y;
```

where "warp" takes proximity into account. The most convenient definition of warp is

```
warp := d/(d − 1 − z);    (*d is the "depth" parameter*)
```

which has the following interpretation. Assume that *one* of your eyes resides at $z = d − 1$, where $d > 1$. Consider every line that goes from your eye through a *rotated point* resulting from the "rotate" procedure above, and plot, on the screen, where these lines strike $z = −1$. This means that a point (x,y,z) that, after rotation, has z greater than $−1$ will have its x and y magnified. The correct factor of magnification turns out to be "warp" as given above. Notice that $z = −1$ causes "warp" to have the value unity. Thus points in the back, that is on the plane $z = −1$, will experience no magnification. Points with $z = +1$ after rotation will be expanded the most.

In the procedure "perplot" in Figure 7.9, "move", "plot", and "draw" include perspective. For example, "perplot (0.3,0.4,0.6,1,2,3,2)" rotates the point $(0.3,0.4,0.6)$ by Euler angles $(1,2,3)$ and draws to this point. This is all done with perspective depth value assigned to "d". An application of this procedure will now be described.

We shall model *enantiomeric forms* of a molecular compound, namely 3-methylhexane, which has two optical isomers. One of these isomers will be

```
   procedure perplot(x, y, z, a, b, c: real; n: integer);
(* plot, move, draw with perspective *)
   var
      warp: real;
   begin
      rotate(x, y, z, a, b, c);
      warp := d / (d - 1 - z); (* this is perspective warp factor *)
      x := x * warp;
      y := y * warp;
      if n = 1 then
         plot(x, y)
      else begin
         if n = 0 then
            move(x, y)
         else
            draw(x, y)
      end
   end; { perplot }
```

Figure 7.9

drawn as if we are looking down the z axis (toward $-z$) and the x axis runs to our east and the y axis to our north. We place groups at vertices of a regular tetrahedron as follows:

$$CH_2CH_3 \text{ at } \left(1, 0, \frac{-1}{\sqrt{8}}\right)$$

$$CH_3 \text{ at } \left(\frac{-1}{2}, \frac{\sqrt{3}}{2}, \frac{-1}{\sqrt{8}}\right)$$

$$CH_2CH_2CH_3 \text{ at } \left(\frac{-1}{2}, \frac{-\sqrt{3}}{2}, \frac{-1}{\sqrt{8}}\right)$$

$$H \text{ at } \left(0, 0, \frac{3\sqrt{2}}{4}\right)$$

The location of the central carbon is

C at $(0,0,0)$

The program "enant" in Figure 7.10 draws this idealized model of the molecule. There are several features of interest:

1. The procedure "perplot" has been incorporated, with a depth of three. This is a shallow depth, and perspective effects should be quite noticeable.
2. *Spokes*, that connect "sites" of the molecule, are drawn with some consideration of the hidden-line technique, although the hidden-line problem is not completely solved in "enant". Connecting lines are correct between their endpoint sites but may crossover sites lying forward, which is not

```
program enant(input, output);
(* display 3-dimensional, perspective figures of enantiomers *)
const
    rhobase = 0.21;             (* base radius of a  nucleus site *)
    d = 3;                      (* this is the perspective depth *)
    scale = 2;                  (* drawing compression factor *)
var
    x, y, z, a, b, c: real;
    xx, yy, zz: array [0..4] of real; (* these will hold tetra coords *)
    j, n: integer;
#include "plibh.i"
#include "plib3.i"

    procedure setup;
(* setup coordinates of nucleus sites *)
    var
       k: integer;
    begin
       xx[0]  := 0;
       yy[0]  := 0;
       zz[0]  := 0;
       xx[1]  := 1;
       yy[1]  := 0;
       zz[1]  := -sqrt(1 / 8);
       xx[2]  := -0.5;
       yy[2]  := -0.87;
       zz[2]  := -sqrt(1 / 8);
       xx[3]  := -0.5;
       yy[3]  := +0.87;
       zz[3]  := -sqrt(1 / 8);
       xx[4]  := 0;
       yy[4]  := 0;
       zz[4]  := 3 * sqrt(2) / 4;
       (* now compress to taste *)
       for k := 0 to 4 do begin
           xx[k]  := xx[k] / scale;
           yy[k]  := yy[k] / scale;
           zz[k]  := zz[k] / scale
       end
    end; { setup }

    procedure site(j: integer);
(* draw in atomic site at j *)
    var
       rho, warp: real;
    begin
       rho := rhobase / scale;
       x := xx[j];
       y := yy[j];
       z := zz[j];
       rotate(x, y, z, a, b, c);
       warp := d / (d - 1 - z);
       cir(x * warp, y * warp, rho * warp)
    end; { site }

    procedure perplot(x, y, z, a, b, c: real; n: integer);
(* plot, move, draw with perspective *)
    var
       warp: real;
    begin
       rotate(x, y, z, a, b, c);
       warp := d / (d - 1 - z);         (* this is perspective warp factor *)
       x := x * warp;
       y := y * warp;
       if n = 1 then
           plot(x, y)
       else begin
           if n = 0 then
               move(x, y)
           else
               draw(x, y)
       end
    end; { perplot }
```

Figure 7.10

```
      function front(j, i: integer): boolean;
(* return false if i site is in front of j site, after (a,b,c) *)
      var
         zcoor, xxx, yyy, zzz: real;
         k: integer;

         procedure equate(k: integer; var xxx, yyy, zzz: real);
         begin
             xxx := xx[k];
             yyy := yy[k];
             zzz := zz[k];
             rotate(xxx, yyy, zzz, a, b, c)
         end; { equate }

      begin
         k := i;
         equate(k, xxx, yyy, zzz);
         zcoor := z;
         k := j;
         equate(k, xxx, yyy, zzz);
         if zzz > zcoor then
             front := true
         else
             front := false
      end; { front }

      procedure connect(i, j: integer);
(* draw intersite lines *)
      var
         g, h: integer;
         warp1, warp2, x1, x2, y1, y2, z1, z2, ff, rho, d12: real;
      begin
         g := i;
         h := j;
         if front(j, i) then begin
             h := i;
             g := j
         end;
         x1 := xx[g];
         x2 := xx[h];
         y1 := yy[g];
         y2 := yy[h];
         z1 := zz[g];
         z2 := zz[h];
         d12 := sqrt(sqr(x1 - x2) + sqr(y1 - y2) + sqr(z1 - z2));
         rho := rhobase / scale;
         ff := 1 - rho / d12;
         (* start a connecting bond at the rearmost site of the two *)
         perplot(x1 + (x2 - x1) * ff, y1 + (y2 - y1) * ff,
                 z1 + (z2 - z1) * ff, a, b, c, 0);
         rotate(x1, y1, z1, a, b, c);
         rotate(x2, y2, z2, a, b, c);
         warp1 := d / (d - 1 - z1);
         (* that was warp factor for front site *)
         warp2 := d / (d - 1 - z2);
         (* that was warp factor for rear site *)
         x1 := x1 * warp1;
         x2 := x2 * warp2;
         y1 := y1 * warp1;
         y2 := y2 * warp2;
         d12 := sqrt(sqr(x2 - x1) + sqr(y2 - y1));
         (* this was the paper distance between site centers *)
         ff := rho / d12 * warp1;
         (* that was factor to end line at visual boundary of near site *)
         draw(x1 + (x2 - x1) * ff, y1 + (y2 - y1) * ff)
      end; { connect }
```

Figure 7.10 (continued)

```
begin
    setup;
    readln(a, b, c);
    graph;
    for j := 0 to 4 do begin
        n := 0;
        perplot(xx[j], yy[j], zz[j], a, b, c, n);
        alphal;
        case j of
            0:
                writeln('C');
            1:
                writeln('H');
            2:
                writeln('CH2CH2CH3');
            3:
                writeln('CH2CH3');
            4:
                writeln('CH3')
        end
    end;
    connect(1, 2);
    connect(2, 3);
    connect(3, 1);
    connect(4, 1);
    connect(4, 2);
    connect(4, 3);
    for j := 1 to 4 do
        connect(0, j);
    for j := 0 to 4 do
        site(j)
end.
```

Figure 7.10 (continued)

desirable, so care must be taken in choice of orientation angles. Likewise, "sites" do not block other "sites" in "enant".

3. The use of the "case" statement allows easy selection of the isomer we wish to model.

If we run the program "enant" and input three Euler angles, the result is a three-dimensional perspective drawing of the molecule (that is, one of the enantiomers) (Figure 7.11). It is possible to get the other one by slight modification of the program (Figure 7.12). See the following exercises.

Exercises

1. Convince yourself that the two enantiomers shown in Figures 7.11 and 7.12 are really different.
2. Write a powerful three-dimensional display program that will handle large molecules, have a clear algorithm for entering what molecule goes where, and correctly handle hidden lines.
3. Try to write a program similar to the one you wrote for exercise 2 to analyze *structural* and *geometric isomers*. Write a program to decide when two molecules are identical (given rough coordinates for each group).
4. Suppose that you have a plotter that can draw with red or blue ink on computer command. (The Hewlett-Packard 7470A is such a plotter, and the procedure "color" of the "plibh.i" include file, Appendix A, handles these

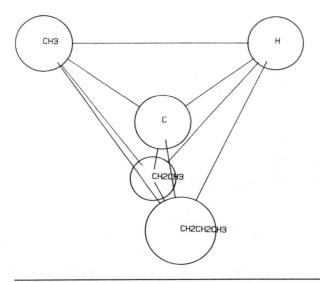

Figure 7.11 Tetrahedral molecule of 3-methylhexane drawn by the program "enant", showing perspective.

pens.) Modify the program "enant" or work out a similar program to draw a double image for viewing through red/blue stereoscopic glasses. You will need to work out the correct algorithm for offsetting the red and blue drawings with respect to each other. *Hint:* Best results for viewing are obtained by using transparency pens on clear plastic sheets and then viewing through red/blue glasses the image projected with a standard overhead projector.

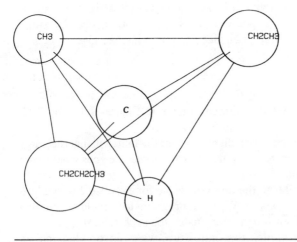

Figure 7.12 Another picture of 3-methylhexane. Is this the optical isomer of Figure 7.11?

Answers

1. They really are different.
2. This exploratory exercise should involve hidden lines, as discussed in Chapter 6.
3. This exercise is exploratory.
4. This exercise is exploratory.

QUANTUM MECHANICAL CALCULATIONS

Nonrelativistic quantum mechanics, as embodied in the *Schrödinger equation,* is thought to account for almost all observed atomic structure. Unfortunately, no system more complicated than neutral hydrogen has ever been solved in closed form. This failure is related to difficulties in the N-body problem of classical mechanics for N greater than 2. Algorithms exist, however, for placing upper and lower bounds on energies (eigenvalues) E for the Schrödinger problem. We will discuss the *Rayleigh-Ritz,* or *variational,* method since it is one of the simplest bounding schemes.

Choose a *trial function* $\psi(\mathbf{R})$, where

$$\mathbf{R} = (\mathbf{r}_1, \mathbf{r}_2, \ldots, \mathbf{r}_N)$$

is a 3N-dimensional collection of real numbers, with each triple of real r_j's denoting the spatial coordinates of particle j, which is usually but not always an electron. The variational result is that the ground-state energy of the Schrödinger system is bounded according to

$$E_0 \le \frac{\int \psi^*(\mathbf{R}) H \psi(\mathbf{R}) \ d^{3N} \mathbf{R}}{\int \psi^*(\mathbf{R}) \psi(\mathbf{R}) \ d^{3N} \mathbf{R}}$$

where the integration is taken over the volume of the 3N space and H is the *Hamiltonian operator* depending on the atomic, ionic, or molecular system under discussion. We will apply this bound formula to the *helium atom* as an example of Pascal programming involving complicated integrals. It is interesting to note that we can write a short, elegant program in spite of the inherent profundity of the quantum mechanical problem.

The atomic Hamiltonian is taken to be

$$H = -\sum_j \nabla_j^2 - \sum_j \frac{Ze^2}{|\mathbf{r}_j|} + \sum_{i \ne j} \frac{e^2}{|\mathbf{r}_i - \mathbf{r}_j|}$$

where j and i have values 1 and 2, respectively,

Z = atomic number of nucleus
e = fundamental electron charge
$1 = \hbar = 2$ m in our units, where m is electron charge

The nucleus is assumed to be infinitely massive. For the helium atom we have $Z = 2$. Assume a trial function of the form

$$\psi(r_1, r_2) = f(r_1)f(r_2)g(z)$$

where

$$r_j = |\mathbf{r}_j|$$

$$z = \frac{\mathbf{r}_1 \cdot \mathbf{r}_2}{r_1 r_2}$$

The variable z is the cosine of the angle subtended by the two electron position vectors. For this class of functions, the Hamiltonian reduces to a simple form. Letting $e = 1$, we have

$$H = -\sum_j \left(\frac{1}{r_j^2} \frac{\partial}{\partial r_j} r_j^2 \frac{\partial}{\partial r_j} + \frac{2}{r_j} \right) - (r_1^{-2} + r_2^{-2}) \frac{\partial}{\partial z} (1 - z^2) \frac{\partial}{\partial z} + \frac{1}{r_{12}}$$

where r_{12}, the distance between electrons, is

$$r_{12} = \text{sqrt}(r_1^2 + r_2^2 - 2r_1 r_2 z)$$

The energy bound can now be given in terms of the functions

$$d \ln f = \frac{f'}{f} = \text{logarithmic derivative of } f$$

$$d \ln g = \frac{g'}{g} = \text{logarithmic derivative of } g$$

as

$$E_0 \leq \frac{\int (\text{integ}) \; dr_1 \; dr_2 \; dz \; (\text{factor})}{\int (\text{integ}) \; dr_1 \; dr_2 \; dz}$$

with

$$\text{integ} = r_1^2 \; r_2^2 \; f^2(r_1)f^2(r_2)g^2(z)$$

$$\text{factor} = [d \ln f(r_1)]^2 + [d \ln f(r_2)]^2 + \frac{1}{r_{12}} - \frac{2}{r_1} - \frac{2}{r_2} + (r_1^{-2} + r_2^{-2})[d \ln g(z)]^2$$

This integral form evolves in a straightforward manner from the restricted hamiltonian and the integration by parts of its Laplacian terms.

The program "helium" in Figure 7.13 computes energy bounds and performs pure Monte Carlo integration. The logic is as follows:

```
program helium(input, output);
(* CALCULATE GROUND ENERGY OF HELIUM USING PURE MONTE CARLO *)
const Z = 2;   (* nuclear charge *)
      k = 15;  (* largest random radius *)
      convert = 54.4; (* of ev's per unit answer in present units *)
var a,b,c,num, den, factor, integ, z, rl, r2, rl2: real;
    ctr: integer;
procedure choose(var rl, r2, z: real);(* get random variables *)
      begin
          rl:=2*random(1)-1; r2:=2*random(1)-1; z:=2*random(1)-1
      end; { choose }
function f(r: real): real; (* radial wave part *)
      begin
          f := exp(-27/32*r);
      end; { f }
function g(z: real): real; (* angular wave part *)
      begin
          g := 1 + a * z + b * z * z + c * z * z * z
      end; { g }
function dlnf(r:real):real; (* derivative of lnf *)
      const ar=0.0001;
      begin
          dlnf := (f(r + ar) - f(r)) / (ar * f(r))
      end; { dlnf }
function dlng(z: real): real;   (* derivative of ln g *)
      const az = 0.001;
      begin
          dlng := (g(z + az) - g(z)) / (az * g(z))
      end; { dlng }
begin
    repeat
        choose(a, b, c);             (* get random angular polynomial *)
        a:=0.2*a; b:=0.2*b; c:=0.2*c;     (* scale coefficients *)
        ctr := 0; num := 0; den := 0;
        repeat
            choose(rl, r2, z);  (* go get random variables *)
            rl := abs(rl) * k; r2 := abs(r2) * k;
            integ := sqr(rl * r2 * f(rl) * f(r2) * g(z));
            rl2 := sqrt(rl * rl + r2 * r2 - 2 * rl * r2 * z);
            factor := sqr(dlnf(rl)) + sqr(dlnf(r2));
            factor := factor +
                sqr(dlng(z))*(1 - z*z) * (1/sqr(rl) + 1/sqr(r2));
            factor := factor + 1/rl2 - Z/rl - Z/r2;
            num := num + factor * integ;
            den := den + integ;
            ctr := ctr + 1
        until ctr = 100000;
        writeln(a:3:3,b:3:3,c:3:3,' E0 =',convert * num / den:4: 4)
    until 0 = 1
end.
```

Figure 7.13

1. Generate a "random" $g(z) = 1 + az + bz^2 + cz^3$, that is, a, b, c are random.
2. Select many points (r_1, r_2, z) and use them to add up the integrands of numerator and denominator integrals.
3. After step 2 is performed a certain number of times, output the E_0 bound.

   In the program "helium", the coefficients a, b, and c of the angular correlation function g are restricted to lie in the range $(-0.2, 0.2)$. The number of Monte Carlo points for step 2 is taken to be 10,000. The particular form for the function f:

$$f(r) = \exp\left(\frac{-27r}{32}\right)$$

```
Variational estimates for helium ground energy, test functions:
   psi = exp(-27(r1+r2)/32) * (1 + az + bz*z + cz*z*z)
where a,b,c are random on (-0.2,0.2)

      a       b       c    electron-volts

   0.133-0.083 0.007 E0 =-72.2350
   0.181 0.060-0.147 E0 =-74.8263
   0.038-0.041 0.120 E0 =-73.9828
   0.120 0.068 0.083 E0 =-72.4101
  -0.166 0.052 0.145 E0 =-75.7224
   0.089-0.035 0.095 E0 =-71.7338
   0.136 0.002-0.112 E0 =-77.5675
   0.052-0.007-0.041 E0 =-78.5953   <--- close to experimental energy
   0.163-0.005 0.157 E0 =-66.7209
   0.155-0.101-0.164 E0 =-75.0914
  -0.166-0.099-0.088 E0 =-74.0294
  -0.068-0.118-0.164 E0 =-74.0229
   0.141 0.081 0.165 E0 =-67.1935
   0.103-0.047-0.052 E0 =-75.5511
   0.107 0.088 0.026 E0 =-74.4647
  -0.015-0.098-0.023 E0 =-75.7960
  -0.143 0.048-0.046 E0 =-74.4415
```

Figure 7.14

arises from the known fact that the best variational bound for helium, in the special case $a = b = c = 0$, has this *screened charge* factor 27/32. This is an analytical result derived in many quantum mechanics textbooks.

The program "helium" is quite simple. For example, step 2 is

num := num + integ*factor;
den := den + integ;

because in terms of digital approximation, integrals are just sums.

The output of the helium calculations is shown in Figure 7.14. Note that only for special choices of a, b, and c is the result close to experimental ionization energy for helium, namely,

E_0(experimental) $= -78.98$(electron-volts)

You should be aware of three points: (1) Monte Carlo procedures are slow; (2) many integrals such as the ones we use in the program "helium" can be done in closed form; (3) variational calculations for helium in particular are by now extremely sharp (Abdel-Raouf, 1982).

Another interesting programming solution for the problems of quantum chemistry is use of the *Hartree self-consistent field* (SCF) *method.* These calculations are based on the notion that one electron of helium should see a *smeared potential* as a result of another electron, and, therefore, we should get two Schrödinger equations, in the SCF approximation, with complicated potential but only one particle in each (Harriss and Rioux, 1980). Solving interactive Schrödinger equations is discussed in Chapter 8, section titled *Quantum Mechanics.*

Exercises

1. Modify the program "helium" in Figure 7.13 to do a multistep Monte Carlo as follows. Run the existing program until you get a list of good triples (a,b,c), for example, those having energy result less than -50. Then increase the iteration number in the "until" statement and from random numbers near these good triples run only for these special numbers, or better. In this way, you will end up with a best Monte Carlo function for $g(z)$. Do not forget the flat function $(a,b,c) = (0,0,0)$.

2. Modify the program "helium" in Figure 7.13 to run with functions of the form

    $$\psi = \exp(-kr_1)\exp(-kr_2)z^{q-1}$$

    Find real pairs (k,q) that give good bounds.

3. Research the problem of doing some of the integrals exactly or by other means. Try to obtain a number that is good to three decimals. This will be the established bound times 26.6 $(= \text{convert}/2)$. The established bound is $-2 \text{ num/den} = 2.903724377\ldots$ (Abdel-Raouf, 1982).

4. Try to extend the methods in exercises 1 to 3 to larger atoms or to higher helium states.

Answers

1. The resulting angular part is almost flat, as evidenced by the fact that the best-known helium wave functions tend to have the expectation $\langle\cos(\text{theta})\rangle$ nearly zero.

2. The best q values will be greater than unity.

3. This exercise is exploratory.

4. This exercise is exploratory.

8 | Examples from Physics

What makes planets go around the sun? At the time of Kepler some people answered this problem by saying that there were angels behind them beating their wings and pushing the planets around in orbit. As you will see, the answer is not very far from the truth. The only difference is that the angels sit in a different direction and their wings push inwards.

R. P. Feynman
The Character of Physical Law (1965)

MECHANICS

Machines are now able to track the meanderings of those "angels" rather well. In fact, pinpoint accuracy of space trajectories, made possible by digital computers, enables space shuttles to touch down within a few meters of their destinations. The machinery, however, has to be programmed to understand the laws of gravity, and one way to do this is to program a differential equation. For example, in Chapter 3 we modeled a projectile in the pull of the earth's gravity. In this chapter we will see how Pascal programs can be used to solve more complex problems and to achieve greater accuracies for simple problems.

A natural starting point for this chapter on physics is the subject of *mechanics.* Consider the *swinging Atwood machine* (SAM), which appears as a double pendulum, with right-hand bob mass = m, left-hand mass = unity, and only the left-hand mass allowed to swing (in two dimensions). The coordinates of the problem, illustrated in Figure 8.1, will be

r = length of string from pulley A to mass 1
th = angle string makes with respect to vertical

We do not need coordinates for the right-hand mass since its position is determined by r together with a constant string length, because of the *no-swing hypothesis* for that mass. The original question asked by Tufillaro (1981), who first posed the SAM problem, is as follows: Given certain masses m *even greater than one,* is it possible that the left-hand mass will achieve a periodic orbit? The supposition is that by virtue of centrifugal forces arising from swinging, mass 1 would effectively

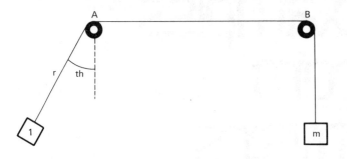

Figure 8.1

balance mass m. This is a difficult question and is ideal for preliminary computer analysis.

Let us digress momentarily to see what kinds of equations we can expect for mechanics problems. Usually it is possible to write down a *Lagrangian* in the following form: (Goldstein, 1980):

$$L = T(q_i, \dot{q}_i) - V(q_i)$$

where

T = kinetic energy as a function of coordinates $q_1, \ldots, q_{dim}$ and velocities $\dot{q}_1, \ldots,$
$\dot{q}_{dim}$
V = potential energy as a function of coordinates $q_1, \ldots, q_{dim}$

and dim represents the total number of space dimensions, and the dot indicates the time derivative. This is not the most general mechanical Lagrangian, but it is general enough to include swinging pendulums, planetary orbits, and any real-world situation in which forces are functions only of position.

For our problem, we have

$$T = \frac{1}{2} m\dot{r}^2 + \frac{1}{2}\dot{r}^2 + \frac{1}{2} r^2\dot{th}^2$$

$$V = -r\cos(th) + mr \text{ (potential energy, with } g = 1)$$

The potential energy V is ambiguous up to an unimportant constant.

The general lagrangian equations of motion are

$$\frac{d}{dt}\left(\frac{\partial L}{\partial \dot{q}_i}\right) - \frac{\partial L}{\partial q_i} = 0 \qquad \text{each } i = 1, \ldots, \text{dim}$$

In the SAM case these reduce to

$$(1 + M)\ddot{r} = r(\dot{th})^2 + \cos(th) - m$$

$$\frac{d}{dt}(r^2\dot{th}) = -r\ \sin(th)$$

We can put the equations of motion thus obtained into a form that will be most useful in Pascal programs. First, we denote notation for generalized position, velocity, and acceleration:

$r, \dot{r}, \ddot{r}$ = pos[1], vel[1], acc[1], respectively
th, th, th = pos[2], vel[2], acc[2], respectively

The equations of motion in Pascal thus become

acc[1] := (pos[1]*sqr(vel[2]) + cos(pos[2]) − m)/(1 + m):
acc[2] := −(sin(pos[2]) + 2*vel[1]*vel[2])/pos[1];

where the second equation is obtained by differentiating the product $r^2\dot{th}$. This result suggests that we can generalize all mechanics problems having the given Langrangian form. The program "solver" in Figure 8.2 acts as a general

```
program solver(input, output);
(* solver for mechanics problems *)
const
     dim = 2;                      (* this is your number of space dimensions *)
     dt = 0.001;                   (* this is your small time increment *)
type
     vector = array [1..dim] of real;
var
     pos, vel, acc: vector;
     t: real;

     procedure readvec(var a: vector);
     var
      i: integer;
     begin
      for i := 1 to dim do
           read(a[i])
     end; { readvec }

     procedure update(var a: vector; b: vector);
     var
      i: integer;
     begin
      for i := 1 to dim do
           a[i] := a[i] + dt * b[i]
     end; { update }
begin
     write('initial pos components: ');
     readvec(pos);
     write('initial vel components: ');
     readvec(vel);
     repeat
      t := t + dt;
      (* insert mechanics formula for 'acc' components here *)
      update(vel, acc);
      update(pos, vel)
     until 0 = 1
     (* do your output here as desired *)
end.
```

Figure 8.2

```
program sam(input, output);
(* sam solver and plotter *)
const
    dim = 2;            (* this is your number of space dimensions *)
    dt = 0.001;              (* this is your small time increment *)
type
    vector = array [1..dim] of real;
var
    pos, vel, acc: vector;
     (* pos[1]=radius, pos[2]=angle, in Lagrangian spirit *)
    t: real;
    ctr: integer;
    m: real; (* the sam non-swing mass *)

#include "plibh.i"

    procedure readvec(var a: vector);
    var
     i: integer;
    begin
     for i := 1 to dim do
         read(a[i])
    end; { readvec }

    procedure update(var a: vector; b: vector);
    var
     i: integer;
    begin
     for i := 1 to dim do
         a[i] := a[i] + dt * b[i]
    end; { update }

begin
    write(' mass ? ');
    readln(m);
    write('initial pos components: ');
    readvec(pos);
    write('initial vel components: ');
    readvec(vel);
    graph;
    move(1, 0.5);
    draw(0, 0.5);
    linetype(6);              (* ready to draw initial string *)
    draw(pos[1] * sin(pos[2]), 0.5 - pos[1] * cos(pos[2]));
    linetype(0);
    repeat
     t := t + dt;
     (* now do the particular sam mechanics ! *)
     acc[1] := (pos[1] * sqr(vel[2]) + cos(pos[2]) - m) / (1 + m);
     acc[2] := -((sin(pos[2]) + 2 * vel[1] * vel[2]) / pos[1]);
     (* we did it...those were the laws of motion *)
     update(vel, acc);
     update(pos, vel);
     if ctr mod 20 = 0 then
         draw(pos[1] * sin(pos[2]), 0.5 - pos[1] * cos(pos[2]));
     ctr := ctr + 1
    until 0 = 1
end.
```

Figure 8.3

mechanics skeleton that we can modify for particular equations of motion. For example, we can add the following to this skeleton:

1. Equations of motion in the main loop (after "repeat") in the form

```
acc[1] := (function of pos[j],vel[j])
acc[2] := (function of pos[j],vel[j])
   ...        (total of "dim" relations of this type)
```

2. Desired time increment "dt" in "const" declarations
3. Desired "dim", which is the number of space dimensions, in declaration
4. Any special output routines, such as graphics
5. Any new parameters, such as m in the SAM problem

Such changes were made in the skeleton program "solver" for the SAM problem and the resulting program, Figure 8.3, is named "sam". The output of the program "sam" is shown in Figure 8.4 for mass $m = 1.14$ and initial coordinates $r(0) = 1$ and $th(0) = 0.3$. The orbit is quasiperiodic and nearly repeats itself, but it is confined to a bounded region of the (r,th) plane.

The program "sam" is a wonderful example of heuristic use of the computer. Such machine computation aided in the discovery of special SAM orbits, for example the *smile,* shown in Figure 8.5, which apparently repeats forever. On the other hand, there exist *teardrops* such as in Figure 8.6, for mass m equal to a magic number, namely, 3.

Exercises

1. Find a mass m and some set of initial conditions for the SAM problem such that the (repeating) trajectory of the unit mass has the topology of a figure 8.
2. Consider the *springy pendulum,* in which a unit mass bob is connected to a *spring.* The top end of the spring is affixed to a ceiling. The potential energy function is (g will be set to 1)

$$V(r,th) = -r\cos(th) + \frac{1}{2}kr^2$$

where k is the spring constant. Study the classical motion for the system, searching for periodic and quasiperiodic trajectories and so on.
3. A planet moves around a very massive star but for some reason the force law is not inverse square but *inverse cube.* The (two-dimensional) motion is due to potential function

$$V(r,th) = -kr^{-2}$$

that is, V involves a constant k and is patently independent of th. Show the truth of the following theorem numerically. For central forces (such as this

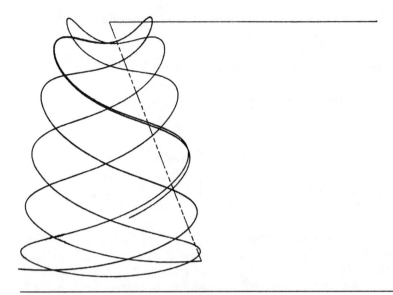

Figure 8.4 Quasiperiodic orbit for a unit-mass bob of a swinging Atwood machine (SAM). The bob starts at the end of the dashed line and apparently never leaves a certain bounded region.

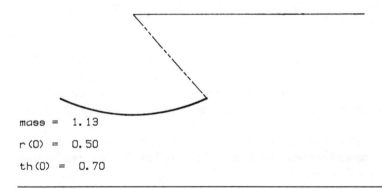

mass =   1. 13

r (0) =   0. 50

th (0) =   0. 70

Figure 8.5

mass =   3. 00

r (0) =   0. 00

th (0) =   1. 00

Figure 8.6

one) the quantity

$$r^2 \dot{th}$$

never changes during the motion. This is the law of conservation of angular momentum, sometimes called *Kepler's Second Law*. Why does the conservation law not apply to the SAM problem in the text?

4. Use Pascal graphics programming to model a *spherical pendulum* whose string has fixed length but whose motion is still two dimensional because of freedom to move on the surface of the sphere defined by radius equal string length. Model the bob as if you are looking down at it. The computer will think the string is actually a rod and will be perfectly willing to let the bob go over the top if the initial conditions are right.

5. Change the programs for some or all of the above exercises so that they run with enhanced accuracy. The *Runge-Kutta-Fehlberg* (RKF) *methods* (Chapter 6 reference: Ralston and Rabinowitz, 1978) are applicable and represent far greater accuracy than the direct iteration we have been using in Chapter 3 and in procedure "advance" in Appendix E.

6. It is said that if we have a pendulum bob of mass m at the end of a massless rigid rod of length L, then by driving the pivot of the rod vertically up and down by $y = y_0 \cos(at)$ there are situations in which the bob stays in a bounded region above the pivot. Is this true?

7. Write a Pascal program to model the motion of a top, with the only constraint that the pivot of the top stay fixed at a point. Exhibit *precession, nutation,* and so on.

8. Suppose that a body moves in two dimensions under the influence of the potential

$$V(r) = r^2 + \frac{h}{r^2}$$

where h is a constant. It is known from theory that the general orbit is a *precessing ellipse.* Verify this with a program, and find values of h such that the orbit has a finite period (finite time to regain its initial conditions).

9. Model a *magnetic monopole* as follows. Put a magnetic charge at the origin of three space so that the magnetic field is

$$B = g \frac{r}{|r|^3}$$

where r is the radius vector. An electric charge q moves according to the *Lorentz force law*

$$F = qv \times B$$

where v is the velocity vector of the charge. For various initial conditions r(0), v(0) of the charge, plot the resulting trajectories. It is known that any trajectory is a *geodesic,* that is, a curve of shortest length, on the surface of some cone. If possible, write a program to find the appropriate cone.

Answers

1. A solution is $m = 1.25$, $r(0) = 0$, $dr/dt(0) = 1$.
2. Solutions similar to the Atwood ones are obtained in the sense that quasiperiodic trajectories result.
3. Every potential $V(r)$ gives a central force, which in turn gives conservation of angular momentum as stated. The SAM does not have this property because the force is not central, since gravity and the string force act together.
4. There are quasiperiodic trajectories.
5. This exercise is exploratory.
6. It is true, as can be shown with Pascal or with a mechanical model. The Mathieu functions play a role in this problem.
7. Precession is the act of the top rolling in an arc around its pivot, nutation is the act of dipping up and down as the precession continues. The Euler equations for rigid bodies, found in Goldstein (1980), are applicable for tops.
8. This exercise is exploratory. It is possible to obtain exact expressions for the orbit, using standard Lagrangian mechanics.
9. This exercise is exploratory.

N-BODY PROBLEMS

Classical mechanics problems having N mass points, to each of which may be attached any number of coordinates, are called *N-body problems*. Beginning with the setting of three mutually gravitating masses, they are generally insolvable in terms of elementary functions. To set up N-body problems we shall need two constants and a time increment. For example,

```
const dim = 2;    (*this is the number of dimensions per particle*)
      num = 800; (*this is the number of particles*)
       dt = 0.001; (*a small time increment*)
```

might be used to model 800 ring particles orbiting the planet Saturn. The library procedure "advance", in Appendix E, is useful for such problems. We can declare

```
type ensemble = array[1..dim,1..num] of real;
```

so that the procedure "advance" can be used. The meaning of the "ensemble" and its associated procedure "advance" is that variables

```
var pos, vel, acc: ensemble;
```

now refer to whole collections of multidimensional coordinates, which procedure "advance" can iterate according to some differential equation. If "pos", "vel", and "acc" refer to ring particles around Saturn,

$vel[2,347]$ = velocity in y direction of particle 347
$pos[1,17]$ = x coordinate of particle 17

and so on. The usual loop will look as follows:

```
repeat
    for n := 1 to num do begin
        for d := 1 to dim do begin
            acc[d,n] := (formula for d-component of acceleration)
        end;
    end;
    advance(vel,acc); (*go and update velocities of all particles*)
    advance(pos,vel); (*likewise update all positions*)
until (condition);
```

This loop has the same effect as the iteration loop of "solver", which was used in the program "sam" (Figure 8.3), except that now all particles numbered "1..num" are involved.

It is worthwhile to contemplate an ensemble as a *gas*. The state of a gas is given classically (when all particles are featureless points) by the collection of all particle positions and velocities. For our Pascal programming purposes, a gas, therefore, is a collection of classical ensembles "pos" and "vel".

The program "saturn" in Figure 8.7 uses "ensembles" to model the following setting:

1. The ring particles initially revolve around Saturn in perfect circular orbits.
2. The ring particles do not interact with each other.
3. A moon of Saturn orbits at fixed radius in steady circular motion. This moon *does* affect the ring particles.

Question: "What happens to the ring because of the moon?" *Answer:* It takes a very long time to answer this question with a program (Crandall, 1978). However, we can answer it as follows. In Figure 8.8, note the plot of just a few ring particles. The initial particle radius at about 0.6 causes the phenomenon of *resonance*, by which the revolving moon perturbs the particle orbit greatly. This is called resonance because it happens when the angular velocity of the particle is an integer multiple of the moon's angular velocity. Effects like this one happen elsewhere in our solar system (Fredrick and Baker, 1976). For the case of Saturn's ring, the computer model explains the *Cassini division,* which is a dark gap in the actual ring (see Figure 8.9). It is apparently caused by Saturn's moon Mimas, which does in fact lie at a radius whose 0.6 multiple is the Cassini radius.

Another application of the type "ensemble" is in the study of *normal modes.* For a number "num" of point masses connected pairwise by springs and obeying the differential equations (recall that acc[i,j] is $d^2pos[i,j]/dt^2$),

$$1 < n < num: \quad acc[1,n] \quad = +k(pos[1, n+1] - 2*pos[1,n] + pos[1, n-1])$$
$$acc[1,num] = +k(pos[1, num-1] - pos[1,num])$$
$$acc[1,1] \quad = +k(pos[1,2] - pos[1,1])$$

where m is a mass constant and k is a spring constant, there will be modes of vibration in which each coordinate "pos[1,n]" oscillates with the same frequency. This *linear coupled oscillator system* can be solved exactly by several methods.

```
program saturn(input, output);
(* track the ring of saturn as perturbed by a moon *)
const
    dim = 2;
    num = 8;
    dt = 0.005;                        (* time increment *)
    a = 0;
    b = 1;
    c = 0;        (* Euler angles for tilted ring *)
    GM = 0.1;     (* this is Universal Gravitational Constant times planet mass *)
    Gm = 0.002; (* this is the same thing but with moon mass *)
(* we demonstrate with phony values *)
type
    ensemble = array [1..dim, 1..num] of real;
(* the ensembles will be pos,vel,acc for ring particles, with
    numbers array[d,1] assigned to the revolving moon *)
var
    pos, vel, acc: ensemble;
    dist, oldpos: ensemble;
    ctr, n, d: integer;
    sr, pr: real;
    pert: real;

#include "plibh.i"
#include "plib3.i"
#include "plibd2.i"

begin
    (* first we place ring and moon *)
    for n := 1 to num do begin
      (* moon first ! *)
      if n = 1 then
          pos[1, n] := 0.8
      else
          (* place a ring particle *)
          pos[1, n] := 0.2 + 0.05 * n;
      (* now compute velocity for circular orbit *)
      vel[2, n] := sqrt(GM / pos[1, n])
    end;
    (* now go into iteration loop *)
    graph;
    cir(0, 0, 0.08); (* this is the planet itself *)
    oldpos := pos;             (* save initial positions for plot *)
    repeat
      for n := 1 to num do begin
          sr := 0;
          pr := 0;
          for d := 1 to dim do begin
              (* get distance in d direction to moon *)
              dist[d, n] := pos[d, 1] - pos[d, n];
              sr := sr + sqr(dist[d, n]);
              pr := pr + sqr(pos[d, n])
          end;
          (* sr is the squared distance to moon *)
          (* pr is the squared distance to saturn *)
          for d := 1 to dim do begin
          if n = 1 then
              pert := 0
          else
              pert := Gm * dist[d, n] / (sr * sqrt(sr));
              (* now do Newton's Law, including moon perturbation *)
              acc[d, n] := -(GM * pos[d, n] / (pr * sqrt(pr))) + pert
      end
    end;
    (* we now have all the accelerations *)
    advance(vel, acc);        (* update velocity ensemble *)
    advance(pos, vel);        (*update position ensemble *)
    if ctr mod 20 = 0 then begin
        for n := 1 to num do begin
            smove(oldpos[1, n], oldpos[2, n], 0, a, b, c);
            sdraw(pos[1, n], pos[2, n], 0, a, b, c)
        end;
        oldpos := pos
    end                       (* save latest plots *);
    ctr := ctr + 1            (* bump display-enable counter *)
  until 0 = 1                        (* or some other condition *)
end.
```

Figure 8.7

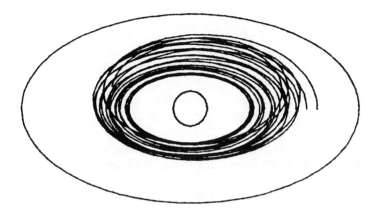

Figure 8.8 Plot of eight particles of a model "ring" of Saturn. The outer orbit is that of a moon whose perturbations on the ring cause precession and general distortion of particle orbits having certain radii. Many more iterations are required to show known features of Saturn.

What is more difficult is the *nonlinearization* of such a problem. A modern example is the *Toda lattice equation,*

$$1 < n < num: \ acc[1,n] = a(exp(b(pos[1,n] - pos[1, n-1]))$$
$$- exp(-b(pos[1, n+1] - pos[1,n])))$$

and the natural generalization of the linear case for the two endpoint equations. This nonlinear lattice has the feature that for ab finite and the limits a → ∞, b → 0, the equations are linear. On the other hand, for ab finite and a → 0 and b → ∞, the limit is the so-called *hard-sphere limit.*

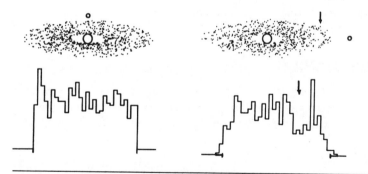

Figure 8.9 A dynamical model of Saturn can be achieved using ensembles of (number) dimension in the hundreds of particles. These before-after plots of the positions of 800 particles plus the moon show definite resonance effect at the Cassini division, which is about 0.6 of the moon's radius. Some, but not many, of the particles have been whiplashed out into space. The histograms show initial random particle placement and final placement with resonance gap.

Exercises

1. Using the type "ensemble", model the motion of a satellite orbiting the earth but at a considerable distance, for example, at 0.2 of the moon's radius. Start the satellite out in a circular orbit and assume the moon does not move (i.e., the earth and the moon sit still in this approximation). Show numerically that the orbit develops a *blister* but that the growth of this disfiguration is roughly perpendicular to the earth-moon line. This sort of behavior is typical of counterintuitive aspects of N-body problems. For example, many people expect the orbit to be elongated toward the moon. Next, install some lunar motion (the right amount for the moon to be moving in a circle perpetually), and investigate whether this makes any significant change in your previous observations.

2. Model the nine planets, without their moons, and attempt to find astronomical data in the literature that are sensitive enough to test for the perturbations predicted by your program. If you wish, model the retrograde motion of Mars as viewed from Earth.

3. Model a tetrahedral arrangement of four masses, each of value m, connected by a total of six springs (one spring per polyhedral edge) each of spring constant k. Use the type "ensemble" to write an elegant loop for getting the four acceleration ensembles. What are the normal mode frequencies?

4. Find solutions to the nonlinear lattice for finite a, b. Graph some traveling waves. This investigation is close to a computer study of *solitons,* discussed later in this chapter.

Answers

1. The orbit bulges at roughly 90 degrees with respect to the earth-moon line. Moon revolution makes little difference until the orbiting radius for the small satellite is greater than about 0.3.

2. This exercise is exploratory.

3. There should be nine normal mode frequencies.

4. This exercise is exploratory.

QUANTUM MECHANICS

Time-independent, or *stationary state,* problems of quantum mechanics lend themselves to be analyzed using Pascal programming. We will begin with the one-dimensional setting in which we must solve for the *wave function* $\psi(x)$ that satisfies

$$-\frac{\hbar^2}{2m}\frac{d^2\psi}{dx^2} + V(x)\psi = E\psi$$

where m is the particle mass, V(x) is the *potential function,* and E is the energy of the state. Recall that this is the time-independent Schrödinger equation. We handle time dependence later in this chapter using very different methods. In what follows, we set

$$\hbar = 1$$

$$m = \frac{1}{2}$$

so that our Schrödinger equation can be rendered in the following shorthand form that is convenient for programming:

$$\psi'' = [V(x) - E]\psi$$

When the wave function $\psi(x)$ has the property that it vanishes for large absolute values of x, we say that ψ represents a *bound state*. The typical example of a bound state is an atomic state whose energy E will change when the electron in a bound state is moved to a different state by absorption or emission of energy. The energy E is called the *bound-state energy*. Usually, bound-state energies occur as discrete values, the idea being that $\psi(x)$ will only vanish at $x = \pm\infty$ for very specific E values. It is this quantization of energies we wish to investigate with Pascal programs.

The first case, and one of the most well-known quantum mechanical problems, is that of the *quantum harmonic oscillator*. Let the potential

$$V(x) = \frac{x^2}{4}$$

represent a *force* linear in x (force is proportional to the derivative of the potential) and therefore a spring potential. It is known that infinitely many bound states exist and that they have energies (Merzbacher, 1970) as follows:

$$E_0, E_1, \ldots, E_n = \frac{1}{2}, \frac{3}{2}, \frac{5}{2}, \ldots, n + \frac{1}{2}, \ldots$$

The state ψ corresponding to energy $n + 1/2$ always crosses the x axis exactly n times. Generally, quantum states are classified in one-dimensional problems as having *even parity* or *odd parity* depending upon whether ψ is an even (symmetric about $x = 0$) or odd function. Therefore, we shall attempt to find numerically some bound states for the quantum oscillator by using the boundary conditions

$$\psi'(0) = 0 \quad \text{and} \quad \psi(0) = 1 \quad \text{the even condition}$$
$$\psi'(0) = 1 \quad \text{and} \quad \psi(0) = 0 \quad \text{the odd condition}$$

Having arranged these initial conditions, the loop we wish to execute is

```
(*denote by "dpsi" the first derivative of psi*)
dpsi := dpsi + (V(x) − E)*psi*dx;
  psi := psi + dpsi*dx;
```

This is the Pascal form of the Schrödinger equation. In addition, we need

1. Loop conditions for exit
2. Some form of output
3. Some form of scaling so output will be reasonably easy to analyze
4. A way of telling if the state is bound

We can now ask whether "psi" blows up to ridiculous values or converges to zero for large abs(x). Solutions to the Schrödinger equation for any smooth *potential* will generally oscillate forever, grow exponentially larger, or damp exponentially, and it is this last decay property that we require for bound states.

The program "qho" in Figure 8.10 plays an interactive game with the user in which the goal is to find an energy-parity combination that gives a convergent plot. When the program is run (on a Tektronix 4012 terminal, easily modified for another device by switching libraries), the prompt for energy and parity, where 1, 0 means odd, even parity, respectively, can be answered as follows:

```
energy   parity
:11.5 1
```

This means we are asking for a plot of the iterated odd parity wave function having energy

$$E = \frac{1}{2} + 11$$

The output in Figure 8.11 shows 11 zero crossings and bound-state behavior in that the wave function tends to zero instead of blowing up as abs(x) increases. It turns out that this ψ will blow up anyway after more plotting of the state. All we can say is that the energy choice 11.5 is *close* to an allowed bound-state energy.

A more complicated setting is the *hydrogen atom*. We have a radial Schrödinger equation

$$-\frac{1}{r^2} \frac{d}{dr} \left(r^2 \frac{d}{dr} \right) \psi + V(r)\psi = E\psi$$

where the potential is the Coulomb function

$$V(r) = \frac{-e^2}{r}$$

We will take the charge e to be $2^{1/2}$, that is, we get a Schrödinger equation

$$\psi'' = -\left(\frac{2}{r} + E \right) \psi - \frac{2}{r} \psi$$

```
program qho(input, output);
(* quantum harmonic oscillator demonstrator *)
(* the game is to try for bound state behavior by choosing
   the correct energy and parity *)
const
    dx = 0.0004;
var
    psiold, xold: real;
    n, f: integer;
    sx, sy: real;
    par: integer;
    x, e, psi, dpsi: real;
    y: real;

    function v(u: real): real;
    begin
     v := 0.25 * u * u
     (* this is the harmonic potential *)
    end; { v }
#include "plibg.i"

begin
    clear;
    write('energy  parity');
    graph;
    move(-1, 0);
    draw(1, 0);
    alpha;
    write('x');
    graph;
    move(0, -1);
    draw(0, 1);
    move(-1.3, 1);
    alpha;
    writeln;
    writeln;
    write(': ');
    readln(e, par);
    x := 0;
    f := 2;
    sx := 0.15 / sqrt(e);
    sy := 0.8 / sqrt(sqrt(e));
    graph;
    repeat
     y := v(x);
     plot(x * sx, y * sy);
     plot(-(x * sx), y * sy);
     x := x + 100 * dx
    until y * sy > 1;
    alpha;
    write('V(x)');
    graph;
    repeat
     graph;
     move(-1.3, 1);
     alpha;
    if f > 2 then begin
        for n := 1 to f do
            writeln;
        write(': ');
        readln(e, par)
    end;
```

Figure 8.10

```
f := succ(f);
sy := 0.8 / sqrt(sqrt(e));
sx := 0.15 / sqrt(e);
if par = 0 then
    psi := sy / 1.5
else
    psi := 0;
if par = 0 then
    dpsi := 0
else
    dpsi := 1;
x := 0;
graph;
xold := 0;
psiold := psi;
repeat
    dpsi := dpsi + (v(x) - e) * psi * dx;
    psi := psi + dx * dpsi;
    x := x + dx;
    if n mod 50 = 0 then begin
        move(xold, psiold);
        draw(x * sx, psi);
        move(-xold, psiold * (1 - 2 * par));
        draw(-(x * sx), psi * (1 - 2 * par));
        xold := x * sx;
        psiold := psi
    end;
    n := n + 1
until (x > 1 / sx) or (abs(psi) > 100000)
    until e = 0
end.
```

Figure 8.10 (continued)

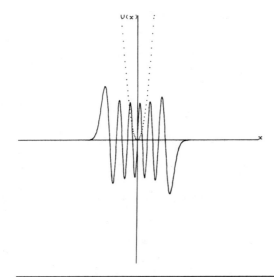

Figure 8.11   The eleventh excited state of the quantum oscillator obtained with program "qho", which solves the Schrödinger equation.

which we shall solve numerically. The particular choices for physical units enable us to observe bound-state energies at

$$E_n = -\frac{1}{(n+1)^2} \qquad n = 0, 1, 2, \ldots$$

The conversion to standard values is obtained by multiplying each of these energies by the actual ground-state energy of hydrogen, about -13.6 electron-volts.

The program "hydro" in Figure 8.12 solves this new Schrödinger equation. However, a trick was invoked to avoid the difficult problems we can encounter near $r = 0$ when potentials are singular. The wave function is substituted according to

$$\phi(x) = \psi(x)\exp(+qr)$$

where q is defined to be sqrt$(-E)$, E always negative. This gives a Schrödinger equation

$$\phi'' = +2q\phi' - \frac{2}{r}(\phi' + \phi - 2q\phi)$$

which turns out to have nice properties near $r = 0$.

Figure 8.13 shows two plots for the energies:

$$E = -0.061$$
$$E = -0.064$$

It is clear from the behavior of the wave functions that there is a bound state with energy lying between these values. In fact, Figure 8.14 shows a successful attempt at energy choice, namely,

$$E = -0.0625 = -\frac{1}{16}$$

This is the fourth bound state and has the correct energy and the required number (three in this case) of zeros.

```
program hydro(input, output);
(* hydrogen atom eigenstate demonstrator *)
const
    dx = 0.01;
var
    nn, q, nnn: real;
    phiold, xold: real;
    n, f: integer;
    sx, sy: real;
    par: integer;
    x, e, phi, dphi: real;
    y: real;

    function v(u: real): real;
    begin
     v := -(2 / u)
    end; { v }

#include "plibg.i"

begin
    clear;
    graph;
    move(-1, 0);
    draw(1, 0);
    alpha;
    write('r');
    graph;
    move(0, -1);
    draw(0, 1);
    move(-1.3, 1);
    alpha;
    write('energy');
    graph;
    x := 0;
    f := 3;
    sx := 0.015;
    sy := 1.5;
    x := dx;
    repeat
     y := v(x);
     if y * sy > -1 then begin
         plot(x * sx, y * sy);
         plot(-(x * sx), y * sy)
     end;
     x := x + 100 * dx
    until x * sx > 1;
    alpha;
    write('V(x)');
    graph;
    repeat
     graph;
     move(-1.3, 1);
     alpha;
     if f > 2 then begin
         for n := 1 to f_do
         writeln;
     write(': ');
     readln(e);
     e := -e;
     q := sqrt(abs(e))
end;
```

Figure 8.12

```
        f := succ(f);
        sy := 1.5;
        sx := 0.015;
        x := 0;
        phi := 1;
        dphi := q - 1;
        dphi := dphi + 2 * q * dphi * dx;
        phi := phi + dx * dphi;
        x := x + dx;
        graph;
        xold := x;
        phiold := phi;
        repeat
            dphi := dphi + (2*q*dphi + 1/x*(-(2*dphi) - 2*phi + 2*phi*q)) * dx;
            phi := phi + dx * dphi;
            x := x + dx;
            if n mod 10 = 0 then begin
                nnn := phiold * exp(-(q * xold));
                nn := phi * exp(-(q * x));
                nnn := nnn * sy;
                nn := nn * sy;
                move(xold * sx, nnn);
                draw(x * sx, nn);
                move(-(sx * xold), nnn);
                draw(-(x * sx), nn);
                xold := x;
                phiold := phi
            end;
            n := n + 1
        until (x > 1 / sx) or (abs(phi) > 100000)
    until e = 0
end.
```

Figure 8.12 (continued)

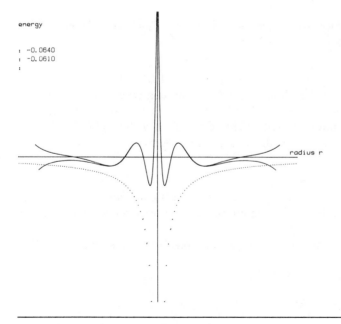

Figure 8.13 Two attempts at guessing the fourth s state of the hydrogen atom. Apparently the correct energy lies between these two energies. This is verified in Figure 8.14.

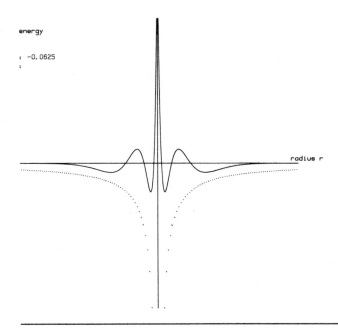

energy

: −0. 0625

radius r

Figure 8.14 Correct guess for the fourth s state of hydrogen, drawn by program "hydro" (Figure 8.12).

Exercises

1. Write a program with which you can interact to obtain an estimate of the ground-state energy for the potential

    $V(x) = abs(x)$

    Compare your results with the known result $E_0 = 1.01879297\ldots$. Then try the same methods for the *quartic potential*, and compare your results with $E_0 = 1.0603620904841820\ldots$ (*Penk's number*). If you are interested in a computer study of power potentials, see Crandall and Reno (1982).

2. Write a program that handles the singular *phenomenological potential*

    $$V(r) = \frac{a}{r} + br$$

    where a is a negative and b is a positive constant. This is a *confining potential*, which can be used to obtain some of the experimental numbers in the theory of *quark binding*.

3. Find the first few states for the *square-well potential* with stump

    $$V(x) = \begin{cases} V_1 & |x| > 0.5 \\ 0 & 0.1 < |x| < 0.5 \\ V_2 & |x| < 0.1 \end{cases}$$

    with V_2 less than V_1. Describe how this function is a model for a molecular ion. Give a physical argument as to why some states have energies so close to another state's energies.

Answers

1. The wave functions are the tails of the Ai(−z) (Airy) functions discussed briefly in Chapter 6. The value for E_0 for V = abs(x) is the first zero of the derivative of Ai(−z).
2. This exercise is exploratory.
3. The physical argument is that the setting is two separated square wells with small correlation (because of the pedestal) between them. Note that two completely separated wells have equal ground- and first-excited states.

SCATTERING

We shall see that *time-dependent* quantum mechanics is far different, at least for programming purposes, than the steady-state analyses of the last section. The time-dependent Schrödinger equation is

$$i\hbar \frac{\partial \psi}{\partial t} = \frac{-\hbar^2}{2m} \frac{\partial^2 \psi}{\partial x^2} + V(x)\psi$$

for one-dimensional motion of a wave function $\psi(x,t)$. This wave function is a function of time and space, as well as being complex-valued.

   The straightforward method of iteration would be to identify

```
const max = 100;
type wave = array[−max..max] of real;
var repsi, impsi: wave;
```

so that "repsi" and "impsi" would be the real and imaginary parts, respectively, of the complex wave function. The various array elements would correspond to the variations of ψ along the x axis. For example,

```
repsi[0]  = real part of ψ(0,t)
impsi[−4] = imaginary part of ψ(−4,t)
```

and so on. Letting $\hbar = 2m = 1$ for convenience, as we did in the last section, the *coupled differential equations* can be written as follows:

$$\dot{R} = -I'' + VI$$
$$\dot{I} = +R'' - VR$$

where the dot refers to time derivative, R = "repsi", and I = "impsi". If we wrote the usual straightforward iteration, as was done in Chapter 3, the first of the above equations would be

```
repsi[j] := repsi[j] + dt*(−impsi[j + 1] + 2*impsi[j] − impsi[j − 1]
            + v(j)*impsi[j])
```

However, it turns out that this method is disastrous for most models of Schrödinger wave scattering. The proper way to modify this relationship is to use *two-time-step methods* (Goldberg et al., 1967).

The program "schro" in Figure 8.15 uses the two-time-step method by first setting up types "wave" and declaring the variables

rpast, rfut, rpres, ipast, ifut, ipres: wave;

This allows for the two-time-step algorithm, which occurs in the double-repeat loop near the end of the main block, as well as for the "frame" procedure in the Appendix E library.

Figure 8.16 shows scattering from a square-well potential. Another good picture is Figure 5.8, which is the special case of scattering when no incident energy is reflected.

The program "schro" can be modified to output the real part of the wave function, which is the way that the bound-state oscillation in Figure 8.17 was generated. Figure 8.17 is a good test of the iteration algorithm, since the periodicity should never falter. A plot of the absolute square of this bound state would show an unchanging rigid distribution centered around $x = 0$.

Exercises

1. Write a program to model scattering of an incident particle by a *lattice*. Put small inverted square-well potentials periodically along a line, and use the incident wave

   repsi = cos(kx) exp($-q(x - a)^2$)
   impsi = sin(kx) exp($-q(x - a)^2$)

   which will have velocity equal to 2k. Study the behavior of overall *throughput* of wave matter versus energy and lattice separation.

2. Look up the exact time-dependent solution for the *free particle* starting with a Gaussian $\omega(x,0)$ and plot it in space-time for a velocity of your choice. Then iterate and increase the number dt until machine error is obvious. This is one way to perform qualitative error analysis of time-dependent problems.

3. Solve the *diffusion equation*

   $$\frac{\partial p(x,t)}{\partial t} = D \frac{\partial^2 p}{\partial x^2}$$

   for constant D using the double-time-step technique. Generally, a starting distribution p(x,0) simply spreads out. Next, solve for diffusion in a box defined by constraints $-a < x < a$, with reflection off the walls, and show that an initial distribution within the walls always goes toward a flat distribution. To model reflection at walls, force the spatial first derivative to be zero at $\pm a$. One way to do this is to erect *image sources,* which are copies of the original source placed at regular intervals along the x axis but outside of the box.

```
program schro(input,output);
(* solves the time-dependent Schroedinger equation using a very
   accurate iteration technique of Golberg, Schey, and Schwartz
   American Journal of Physics, Vol. 35, No. 3 (1967) *)
(* this program is by B. R. Litt *)
const
    max = 100;
    m = 40;   (* number of iterations per picture *)
    dt = 0.0025;
type
    wave = array [-max..max] of real;
var
    vscale,yscale:real;   (* scaledown factors for pictures *)
    vv,LL:real;
    ii, j, l: integer;
    temp, dx: real;
    psisq, va, rpres, ipres, ipast, rpast, ifut, rfut: wave;
    a, k, s, d, ao: real;

# include "plibh.i"  (* include HP7470 graphics library *)
# include "plib3.i"  (* include 3d library *)
# include "plibdl.i"  (* include dynamical models library part 1 *)
    function shape(j:integer):real;
    (* initial shape psi(x,0) to be evolved *)
    begin
        shape := ao * exp(-(sqr(j + a) / (4 * sqr(s))))/yscale;
    end; { shape }

    function v(j: integer): real;
    (* the scattering potential *)
    begin
        v := -(4.0 * d * sqr(1 / (exp(j * dx) + exp(-(j * dx)))))
        {v(j*dx)}
    end; { v }

    procedure setup(var rpres,ipres:wave);
    (* setup real and imaginary parts of initial wave function *)
    var
        ii: integer;
    begin
        for ii := -max to max do begin
            rpres[ii] := shape(ii) * cos(k * ii * dx);
            ipres[ii] := shape(ii) * sin(k * ii * dx)
        end
    end; { setup }

begin
    read(d);   (* get coefficient in v(x) = -d sech^2 x *)
    readln(k);   (* mean momentum of incident packet *)
    s := 9.0;                      (* spread in x, sigma *)
    dx := 0.1;
    ao := 1 / (1.5832335 * sqrt(s * dx));        (* normalization factor *)
    yscale := 0.4 / (ao * ao);
        vscale:= d*5;
        a:= 35;
        setup(rpres,ipres);
        clear;graph;ax;
        (* let's go and draw in v(x) once *)
        for ii:= -max to max do psisq[ii]:=v(ii)/vscale;
        frame(psisq,0);
        (* that just drew it *)
    LL := dt / (dx * dx);
    l := 1;
    rpast := rpres;
    ipast := ipres;
    (* now compute one iteration using crude Euler method, to
                        prepare for use of two time levels *)
    for j := -max + 1 to max - 1 do begin
        vv := v(j);
        rpres[j] := rpres[j] - LL * (ipres[j + 1] + ipres[j - 1]) +
                    (2 * LL + vv * dt) * ipres[j];
        ipres[j] := ipres[j] + LL * (rpres[j + 1] + rpres[j - 1]) -
                    (2 * LL + vv * dt) * rpres[j]
    end;
```

Figure 8.15

```
LL := 2 * dt / (dx * dx);
for j := -max to max do
    va[j] := 2 * (dt * v(j) + LL);
repeat
    repeat
        for j := -max + 1 to max - 1 do begin
            rfut[j] := rpast[j] -
                (LL * (ipres[j + 1] + ipres[j - 1]) - va[j] * ipres[j]);
            ifut[j] := ipast[j] +
                (LL * (rpres[j + 1] + rpres[j - 1]) - va[j] * rpres[j])
        end;
        l := l + 1;
        rpast := rpres;
        ipast := ipres;
        rpres := rfut;
        ipres := ifut
    until l mod m = 0;
        for ii := -max to max do begin
            temp := sqr(rpres[ii]) + sqr(ipres[ii]);
            psisq[ii]:=temp;
        end;
            frame(psisq,l*dt/1.5);
    until l*dt > 3;
end.
```

Figure 8.15 (continued)

4. Show with a Pascal program that a gaussian initial condition psi(x,0) can actually slosh back and forth without changing shape for harmonic potential $V(x) = kx^2/2$. The Gaussian should be started with an offset in the form $psi(x,0) = \exp[-(x - s)^2/4b^2]$.

Answers

1. This exercise is exploratory.
2. The initial gaussian packet should spread out (in accordance with the

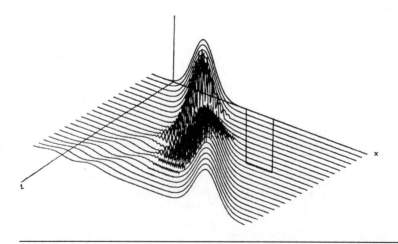

Figure 8.16 Gaussian wave packet scattering from a square-well potential, drawn with a program adapted from "schro" (Figure 8.15). Typical exotic behavior during collision is evident. Courtesy of B. Litt, thesis, Reed College, Portland, Oreg., 1981.

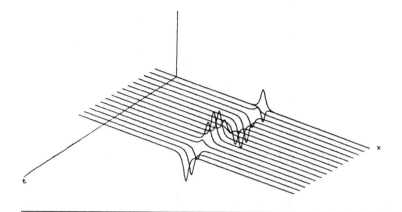

Figure 8.17 Plot of real part of amplitude for the stationary state psi(x,0) = sech x, in potential V(x) = −2 sech² x. Courtesy of B. Litt, thesis, Reed College, Portland, Oreg., 1981.

uncertainty principle) while damping out as 1/sqrt(t) in probability at packet center.

3. The diffusion in the box must go to a flat solution, which looks like a drop of ink diffusing uniformly in a glass of water.

4. The famous result from quantum theory is that the Gaussian does slosh back and forth and, in fact, never changes shape at all (as if it were a cardboard cutout).

SOLITONS

Another application of space-time graphics in physics is the study of *solitons*. These are special solutions to nonlinear differential equations and have the remarkable property of sustaining collision with other solitons. It is interesting that they were discovered with computer programs (Lamb, 1980).

An equation that constantly appears in soliton theory is the *Kortweg-de Vries* (K dV) equation

$$\frac{\partial U}{\partial t} = -U \frac{\partial U}{\partial x} - \frac{\partial^3 U}{\partial x^3}$$

where U is a real function U(x,t). This is nonlinear because of the cross term involving U and its first derivative. The sum of two exact solutions is not generally a solution. An example of a soliton solution is the one plotted by the program "soliton" in Figure 8.18. Figure 8.19 shows a double-soliton event where the faster soliton literally passes through the slower soliton.

The program "soliton" was derived from the skeleton program "spacetime" in Appendix E. Though the program plots only a known function that solves the K dV equation above, you can always write an interaction loop for the program and insert the "frame" procedure in the loop to get the usual running-wave graphics output. The K dV equation is ideal for this task, since it is first order in time.

```
program soliton(input,output);
(* Display a Korteweg-deVries soliton using skeleton
   program "spacetime".  Program by S. Swanson *)

const
        max=100;
        dt=0.1;
        xscale = 10;      (* Solution interesting on [-10,10] *)
type
        wave=array[-max..max] of real;
var
        psi : wave;
        realt, t, psiscale : real;

# include "plibh.i"   (* include plotter library *)
# include "plib3.i"   (* include 3-dimensional library *)
# include "plibdl.i"  (* include dynamical proc's 'ax' and 'frame' only *)
# include "plibl.i"   (* include math library *)

        procedure setup(var psi : wave; t : real);
        (* do proper psi vs. x solution at current time *)
        var ii:integer;
            nx, uu : real;
        begin
            for ii:= -max to max do begin
                  nx := ii /xscale;
                  uu := 4*cosh(2*nx-8*t)+cosh(4*nx-64*t)+3;
                  uu := uu/sqr(3*cosh(nx-28*t)+0.01+cosh(3*nx-36*t));
                  uu := -12 * uu / psiscale;
                  psi[ii] := uu
            end
        end;

begin
        write('Psiscale: ');      (* let user select +/- and scale *)
        readln(psiscale);
        graph;
        ax;                       (* draw axis *)
        t := 0;                   (* t -> (0,2) *)
        realt := -0.4;            (* realt on (-.4,.4) most interesting area *)
        setup(psi, realt);
        frame(psi, 0);
        repeat
                t := t + dt;
                realt := realt + dt / 2.5;
                setup(psi, realt);
                frame(psi, t);
        until t > 1.95;
        alpha
end.
```

Figure 8.18

Another equation, which has certain advantages over the K dV equation, is the *Boussinesq equation:*

$$\frac{\partial^2 U}{\partial t^2} = \frac{\partial^2 U}{\partial x^2} + 6\frac{\partial^2(U^2)}{\partial x^2} + \frac{\partial^4 U}{\partial x^4}$$

A program similar to "soliton", with the modification that an iteration loop replaced the direct function evaluation, was run for the Boussinesq equation. The result is that a starting function U(x,0), actually a Gaussian, turned into three solitons. See Figure 8.20.

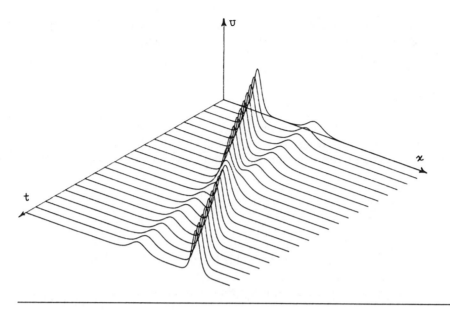

Figure 8.19 A famous double-soliton solution of the Kortweg-de Vries equation. The faster (taller) soliton passes through the slower (smaller) one, with both intact after the collision.

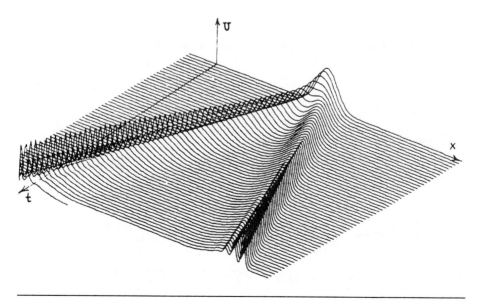

Figure 8.20 This Gaussian initial condition for the Boussinesq equation eventually turns into three little solitons in each direction of propagation.

Solitons are important in various branches of physics, especially the study of magnetic domains, nonlinear biophysical systems, low-temperature physics, and particle physics.

Exercises

1. By trying initial wave distributions $U(x,0)$ and iterating the KdV differential equation, find a $U(x,0)$ that disintegrates into three solitons.
2. Write a program to display numerical solutions to the nonlinear Schrödinger equation

$$i\frac{\partial U}{\partial t} = -\frac{\partial^2 U}{\partial x^2} - KUU^*U$$

where $U(x,t)$ is generally complex-valued. This describes the self-focusing of a light wave in certain media.
3. Investigate with Pascal programs the *sine-Gordon equation,*

$$\frac{\partial^2 U}{\partial t^2} = \frac{\partial^2 U}{\partial x^2} - \sin(U)$$

which is known to have soliton solutions in the form of *kinks* and *antikinks.* These will be apparent if you plot, for your vertical space-time axis, the quantity U' instead of U.

Answers

1. This is equivalent to finding a potential V in Schrödinger theory having exactly three bound states. Thus, for example, there will be a function of the form

   U(x,0) = −d/sqr(cosh(x))

   having the required property (some d).
2. This exercise is exploratory.
3. An exact solution is

   U(x,t) = 4*arctan(exp(−(x − vt)/sqrt(1 − sqr(v))))

   which moves with speed v.

9 | Examples from Biology

Serious students of cellular and developmental biology confront what may be the gravest epistemological problems ever faced by scientists. These are a direct consequence of the immense, ordered biochemical complexity of organisms.

S. Kauffman
Towards a Theoretical Biology (1972)

POPULATION GENETICS

We hope that computers can be used to solve some of the problems of biology, including the epistemological problems referred to by Kauffman (1972). At present we can model the simple, well-understood biological processes with computers. An example for the moderately experienced Pascal programmer is that of modeling genetic combination and reproductive processes. This will be done in accordance with the microscopic reaction model in Chapter 7, in which individual entities are probabilistically combined.

For our example, you can declare

```
type allele = (a,b);   (*this will be a two-allele system at first*)
     frequency = array[allele] of real;   (*these will be probabilities*)
     genotype = array[allele,allele] of integer;   (*these will be counts*)
var g: genotype; i, j: allele; p: frequency;
```

The var "g" is a genotype count in the sense that

g[i,j] = total count of genotypes i, j

The count is an integer. Note that i, j are not integers—they are *symbols* (a or b in this case) taken from a *set* of alleles. In a given generation, we assume that the alleles are mixed and that the probability of forming genotypes is as follows:

probability of forming genotype ij = p[i]*p[j]

```
program drift(input, output);
(* simple genetic model for drift, with two alleles *)
const lim = 22; (* there will be 22 generations listed *)
type allele = (a, b);
     genotype = array [allele, allele] of integer;
     frequency = array [allele] of real;
var i, j: allele;
    g: genotype;
    p: frequency;
    ctr, generation, total: integer;
begin
    readln(g[a, a], g[a, b], g[b, b]);
    total := g[a, a] + g[a, b] + g[b, b];    (* this will not change *)
    generation := 1;
    repeat
    p[a] := (2 * g[a, a] + g[a, b]) / (2 * total);
    p[b] := 1 - p[a];
    for i := a to b do
        for j := a to b do
            g[i, j] := 0;
    for ctr := 1 to total do begin
        if random(1) < p[a] then i := a
                           else i := b;
        if random(1) < p[a] then j := a
                           else j := b;
        g[i, j] := g[i, j] + 1
    end;
    g[a, b] := g[a, b] + g[b, a]; (* ab and ba are equivalent *)
    writeln(generation, g[a, a], g[a, b], g[b, b]);
    generation := generation + 1
    until generation > lim
end.
```

Figure 9.1

where the probabilities are based on the simplest possible assumption that allele frequency is the total incidence of that allele divided by total alleles, or

$$p[a] = (2*g[a,a] + g[a,b])/(2*total); \quad p[b] := 1 - p[a];$$

with "total" equal to $g[a,a] + g[a,b] + g[b,b]$. The factors "2" come from counting the alleles, for example, there are two "a" alleles in the "a, a" genotype.

The program "drift" in Figure 9.1 models *successive generations*. The program is called "drift" because it exhibits, for small input genotype counts, the phenomenon of *genetic drift*. In other words, it is possible that one allele becomes extinct. For example, the input

3 3 3   (*note that alleles a, b start out on equal footing*)

on one run gives the results shown in Figure 9.2. Note that "b, b" as well as "a, b" genotypes became extinct, that is, there are no more "b" alleles.

If the number of input genotypes is sufficiently large, e.g., in the thousands, a qualitatively different situation arises. Even though there is still a nonzero extinction probability for a given allele, this probability will be extremely small. The input

2000 2000 2000

for example, gives the output shown in Figure 9.3. Note that except for small fluctuations, there is a *quasiequilibrium* reached. This condition will always occur on the basis of the following relationships, which are approximately true for very large genotype counts:

g[i,i] $\cong$ p[i]p[i]*total;   i = a, b
g[a,b] $\cong$ 2*p[a]p[b]*total

Here g denotes the genotype count for the next generation on the basis of the total for the last generation. Whenever p[a], p[b] are computed, it turns out that the above relationships for g are preserved. Therefore if p[a] = p[b] initially, then on each generation, starting with the first pass,

2g[a,a] $\cong$ g[a,b] $\cong$ 2 g[b,b]

which is verified in Figure 9.3. It is interesting that the quasiequilibrium is essentially reached on the first pass of the probability loop.

   The quasiequilibrium is a special example of the *Hardy-Weinberg equilibrium* that generally occurs on the basis of the following assumptions (Roughgarden, 1979):

1.   Natural selection and other frequency-changing forces are absent.
2.   Union of random mating types is uniform (there is no common *gamete* pool).
3.   Reproduction is asexual.

generation	g[a,a]	g[a,b]	g[b,b]
1	1	5	3
2	0	5	4
3	1	2	6
4	2	3	4
5	0	5	4
6	1	5	3
7	1	7	1
8	1	7	1
9	1	5	3
10	3	2	4
11	1	5	3
12	2	5	2
13	1	6	2
14	1	6	2
15	4	5	0
16	8	0	1
17	7	2	0
18	7	2	0
19	7	2	0
20	9	0	0
21	9	0	0
22	9	0	0

Figure 9.2

generation	g[a,a]	g[a,b]	g[b,b]
1	153 8	297 8	14 84
2	1522	3011	1467
3	1492	3010	1498
4	1420	3027	1553
5	1449	3013	153 8
6	1460	2971	156 9
7	1430	3006	156 4
8	1458	2965	1577
9	1467	29 88	1545
10	1449	3005	1546
11	14 88	3021	1491
12	1529	3028	1443
13	1537	3010	1453
14	1552	2992	1456
15	1595	2965	1440
16	1560	2966	1474
17	153 8	3006	1456
18	1527	29 80	1493
19	14 86	3053	1461
20	153 8	301 8	1444
21	1542	3004	1454
22	1506	3020	1474

Figure 9.3

The Hardy-Weinberg equilibrium also occurs in more general settings. When *selection* is involved, for example, the qualitative results change drastically. We will introduce the notion of *fitness* in order to include the effects of selection. We will take a census of the population at the *zygotic* phase, at which time the random union of gametes has created the numbers g[i, j] in accordance with the three equilibrium relationships above (recall that for large numbers of alleles the Hardy-Weinberg law applies after one generation). In the adult phase, each of the three genotype counts must be multiplied by factors L[a,a], L[a,b], L[b,b], which embody

L[i, j] = probability genotype i, j suvives to reproduce

These numbers must then be multiplied by *birth rates* because we want to model fitness as a result of differential birth and survival rates. The fitness numbers will therefore be grouped into constants as follows:

w[i, j] = absolute selective value of genotype i, j, or fitness number for genotype i, j

With several other assumptions, which we do not discuss here, we arrive at the following equation for the rate of change of the allele frequency obtained from simple counting arguments:

$$\frac{dp[a]}{dt} = \frac{p[a]p[b](p[a](w[a,a] - w[a,b]) + p[b](w[a,b] - w[b,b]))}{p[a]p[a]w[a,a] + 2p[a]p[b]w[a,b] + p[b]p[b]w[b,b]}$$

The same equation is true for dp[b]/dt but with (a,b) reversed in all terms. We have already shown possible Pascal-type assignments for genetic combination problems. Therefore we will write our next program without special type declarations so that we can concentrate on the fitness model.

The program "evolve" in Figure 9.4 contains the above equation for rate of change in terms of fitness numbers. The alleles are no longer scalars. Instead we assign

$$p1 = p[a]$$
$$p2 = p[b]$$
$$dpj = \text{time derivative of } p_j$$

```
program evolve(input, output);
(* model the effect of fitness on evolution of allele freq. *)
const
    genscale = 100;
var
    p1, p2, plsav: real;
    wbar: real;
    dp1: real;
    s, w11, w12, w22: real;
    gen: integer;
#include "plibh.i"
begin
    write('frequency pl (p2:= 1-pl) ');
    readln(plsav);
    graph;
    move(1, -0.8);
    draw(-1, -0.8);
    draw(-1, 1);
    move(-1.3, 0);
    alphal;
    writeln('freq. pl');
    move(-0.5, -1);
    alphal;
    writeln('generations');
    w22 := 1;                       (* selection against the dominant allele 'l' *)
    s := 0;
    repeat
    s := s + 0.05;
    w11 := 1 - s;                   (* s is the 'selection coefficient' *)
    w12 := 1 - s;
    pl := plsav;                    (* initialize the frequencies *)
    p2 := 1 - pl;
    gen := 1;
    repeat
        dpl := pl * p2 * (pl * (w11 - w12) + p2 * (w12 - w22));
        wbar := sqr(pl) * w11 + 2 * pl * p2 * w12 + sqr(p2) * w22;
        dpl := dpl / wbar;
        pl := pl + dpl;
        p2 := 1 - pl;
        if gen mod 5 = 1 then begin
            if gen = 1 then
                move(gen / genscale - 1, -0.8 + 1.8 * pl)
            else
                draw(gen / genscale - 1, -0.8 + 1.8 * pl)
        end;
        gen := gen + 1
    until gen > 2 * genscale
    until s > 0.4
end.
```

Figure 9.4

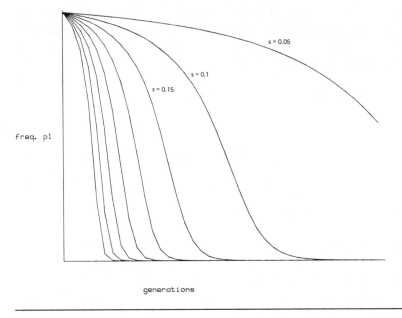

Figure 9.5 Selection against dominant allele 1 showing degradation of allele frequency p1. Plot is from program "evolve" (Figure 9.4), with curves drawn for s in steps of 0.05.

The only input number is p1 at generation zero, after which the calculations are performed for selection against the dominant allele as follows:

$$w_{22} = 1$$
$$w_{11} = 1 - s$$
$$w_{12} = 1 - s$$

where s is a *selection coefficient* for the problem. The curves for various s's are shown in Figure 9.5.

Exercises

1. With a program similar to the program "evolve" in Figure 9.4 (you can use printouts instead of graphics), work out results for the following situations:
    (a) Selection against the recessive allele: $w_{11} = 1 - s$; $w_{12} = 1$; $w_{22} = 1$
    (b) Selection against the homozygotes: $w_{12} = 1$; $w_{11} = 1 - s_1$; $w_{22} = 1 - s_2$
    (c) Selection against the heterozygote: $w_{11} = 1 + s_1$; $w_{22} = 1 + s_2$; $w_{12} = 1$

    In cases (a) and (b), what can you say about the state of affairs after many generations in terms of the numbers s_1, s_2?
2. Modify the program "drift" in Figure 9.1 to handle more than two alleles. One feature of using clever declarations is that sometimes modifications can be performed with minimal effort, that is, by changing only one or two declarations.

3. Consider a drift model (that is, low genotype counts) together with the fitness model (presence of the w_{ij}). Operationally define the relevance of selection as the proportional reduction in extinction time for a given allele, that is, create the new parameter

$$E_j = \frac{\text{extinction time with fitness included}}{\text{extinction time without fitness}}$$

Study this parameter and how it relates to the coefficients w_{ij}.

4. Work out and solve a different sort of problem in population biology. Topics of computer interest follow (Stearns and Crandall, 1981a; Crandall and Stearns, 1982; Stearns and Crandall, 1981b):
   (a) Manipulation of *Leslie matrices*
   (b) Solution of the *Euler-Lotka implicit integral formula*
   (c) Problems of *bet hedging* and *persistence*

Answers

1. For selection against homozygotes, allele 1 frequency approaches the asymptote

   $$p = s2/(s1 + s2)$$

   Heterozygote superiority leads to polymorphism. For selection against heterozygotes, the situation is qualitatively different. If initial allele 1 frequency is less than 1/2, allele 1 reaches extinction; whereas if initial allele 1 frequency is greater than 1/2, allele 2 reaches extinction.
2. This exercise is exploratory.
3. This exercise is exploratory.
4. This exercise is exploratory.

COMMUNITY ECOLOGY

Assume that a population grows at a steady rate r. This means that if $N(t)$ is the time-dependent population density, then

$$\frac{dN}{dt} = rN$$

The solution to this growth equation is the well-known exponential

$$N(t) = N(0) \exp(rt)$$

where $N(0)$ is the population at time zero. Needless to say, the assumption of steady growth rate is not realistic. One way to create a more practical model is to invoke *density dependence.* The *logistic equation* is a simple model for naturally constrained growth and takes the form

$$\frac{dN}{dt} = \frac{rN(k-N)}{k}$$

where k is a new parameter, called the *carrying capacity,* whose mathematical importance will be seen shortly. Note that when the density is quite small and N is much less than k, we have

$$\frac{dN}{dt} \approx rN \qquad N \ll k$$

But if N is nearly equal to k we have

$$\frac{dN}{dt} \approx 0 \qquad N \approx k$$

We interpret k as the *maximum allowed density.* The organism can grow essentially unimpeded when N is small but must stop growing as N approaches the maximum k. An exact solution to the above logistic equation is

$$N(t) = \frac{k \exp(rt)}{k/N(0) - 1 + \exp(rt)}$$

which has the properties claimed.

The situation is much more complicated when competition between species is invoked as a secondary population-limiting process. A two-component model of the growth of two species of respective densities $N_1(t)$ and $N_2(t)$ is

$$\frac{dN_1}{dt} = \frac{r_1 N_1(k - N_1 - c_{12}N_2)}{k}$$

$$\frac{dN_2}{dt} = \frac{r_2 N_2(k - N_2 - c_{21}N_1)}{k}$$

where the competition numbers c_{12} and c_{21} describe the attenuation of growth caused by the competitor. This two-component logistic is solved numerically by the program "phaseplot" in Figure 9.6, which draws a trajectory in (N_1, N_2) space, always starting with the point (0.5,0.5). The method of computation is best understood using the program "ecology" in Figure 9.8, which we describe shortly. For the moment, we can see that the output of the program "phaseplot" (Figure 9.7) reveals a tendency for one or the other species to take over the environment completely. The coefficients c_{12}, c_{21} are taken randomly in the program "phaseplot", and after 10 trial runs, each starting at (0.5, 0.5), there is exactly one run that *stabilizes* at about (0.54,0.62) in this *phase space.* Such stability arises from accidental equality of two particular expressions. If N_1 and N_2 happen to satisfy

$$N_1 + c_{12}N_2 = N_2 + c_{21}N_1 = k$$

```
program phaseplot(input, output);
(* evolve a community of species according to competition logistics *)
const
    species = 2;
    dt = 1.0;
type
    vector = array [1..species] of real;
    matrix = array [1..species, 1..species] of real;
var
    community: vector;       (* this is the set of densities, one per species *)
    threat: vector; (* contains species-wise threat numbers *)
    velocity: vector;        (* this is rate of change of densities *)
    rate: vector;    (* this is the set of growth rates, one per species *)
    capacity: vector;        (* this is the carrying capacity, one per species *)
    competition: matrix;(* this is the competition matrix *)
    t, cc: real;
    i, j: integer;

#include "plibm.i"
#include "plibh.i"

    function extinction: boolean;
    var
    k: integer;
    alive: boolean;
    begin
    alive := true;
    k := 0;
    while alive and (k < species) do begin
        k := k + 1;
        if community[k] < 0.01 then
            alive := false
    end;
    if not alive then
        extinction := true
    else
        extinction := false
    end; { extinction }

begin
    (* first let us fill in rates *)
    for i := 1 to species do
    rate[i] := 0.03;
    (* next we fill in capacities *)
    writeln('capacity constant: ');
    readln(cc);
    for i := 1 to species do
    capacity[i] := cc;
    (* now we get some initial densities *)
    graph;
    plot(1, -0.8);
    draw(-1, -0.8);
    draw(-1, 1);
    move(-1.3, 0);
    alpha1;
    writeln('species 1');
    move(-0.5, -1);
    alpha1;
    writeln('species 2');
    repeat
    community[1] := 0.2;
    community[2] := 0.7;
    (* now make a new competition matrix *)
    for i := 1 to species do begin
        for j := 1 to species do begin
            if i = j then
                competition[i, i] := 1
            else
                competition[i, j] := 3 * random(1)
        end
    end;
    plot(2 * community[1] - 1, 1.8 * community[2] - 0.8);
    (* now we go into the ecological loop *)

    t := 0;
    repeat
        t := t + dt;
        mvprod(species, species, competition, community, threat);
        (* this gets 'threat' as product of 'competition' & 'community' *)
        for i := 1 to species do
            velocity[i] := rate[i] * community[i] *
                            (1 - threat[i] / capacity[i]);
        for i := 1 to species do
            community[i] := community[i] + velocity[i] * dt;
        draw(2 * community[1] - 1, 1.8 * community[2] - 0.8)
    until (t > 300) or extinction
    until 0 = 1
end.
```

Figure 9.6

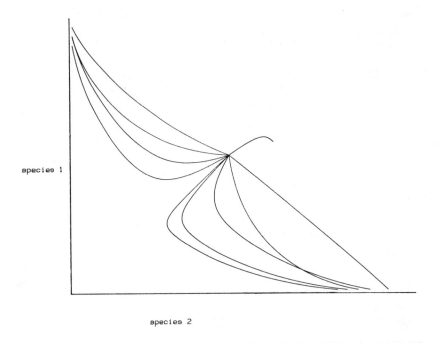

species 1

species 2

Figure 9.7 Phase plot of population densities for two species in competition, made by program "phaseplot" in Figure 9.6. Stability is rare; the usual outcome is extinction of a species.

then the whole growth process comes to a halt: on the phase plot the trajectory terminates at a point (N_1, N_2) rather than at the point $(0, N_2)$ or $(N_1, 0)$.

In Pascal programs, it is often useful to define functions that have no arguments, such as the function "extinction" in program "phaseplot". Note the loop

```
while alive and (k < species) do begin
    k := k + 1;
    if community[k] < 0.01 then alive := false;
end;
```

Here, community [k] is just the density $N(k)$ from the general program "ecology" (Figure 9.8). The loop does one of the following:

1. Exits with "alive" equal to boolean value "true", meaning all densities are greater than or equal to 0.01
2. Exits with "alive" equal to the boolean value "false", meaning at least one species is extinct (in the sense that density is less than 0.01)

If we need to know which species has become extinct, the function could be made to return an integer, with the value "0" returned for no extinct species.

```
program ecology(input, output);
(* evolve a community of species according to competition logistics *)
const
    species = 11;
    dt = 1.0;
type
    vector = array [1..species] of real;
    matrix = array [1..species, 1..species] of real;
var
    community: vector;        (* this is the set of densities, one per species *)
    threat: vector;    (* contains species-wise threat numbers *)
    velocity: vector;        (* this is rate of change of densities *)
    rate: vector;    (* this is the set of growth rates, one per species *)
    capacity: vector;        (* this is the carrying capacity, one per species *)
    competition: matrix;(* this is the competition matrix *)
    t, cc: real;
    i, j: integer;

#include "plibm.i"
begin
    (* first let us fill in rates *)
    for i := 1 to species do
    rate[i] := 1;
    (* next we fill in capacities *)
    writeln('capacity constant: ');
    readln(cc);
    for i := 1 to species do
    capacity[i] := cc;
    (* next let us do competition matrix *)
    for i := 1 to species do begin
    for j := 1 to species do begin
        (* note we always need diagonal as 1's *)
        if i = j then
            competition[i, i] := 1
        else
            competition[i, j] := 0.3 * random(1)
    end
    end;
    writeln('competition matrix: ');
    writemat(species, species, competition);
    writeln;
    (* now we get some initial densities *)
    for i := 1 to species do
    community[i] := 0.5;
    writeln;
    write('time      s1      s2      s3      s4      s5      ');
    writeln('s6      s7      s8      s9      s10      s11');
    writeln;
    (* now we go into the ecological loop *)

    t := 0;
    repeat
    t := t + dt;
    mvprod(species, species, competition, community, threat);
    (* this gets 'threat' as product of 'competition' & 'community *)
    for i := 1 to species do
        velocity[i] := rate[i] * community[i] *
                            (1 - threat[i] / capacity[i]);
    for i := 1 to species do
        community[i] := community[i] + velocity[i] * dt;
    write(t: 3: 3);
    writevec(species, community)
    until 0 = 1
end.
```

Figure 9.8

A general logistic model for arbitrarily many species can be expressed in the form of the vector equation

$$\frac{dN[i]}{dt} = r[i]N[i] \left(1 - \frac{CN[i]}{k[i]} \right)$$

where i represents integers "1 to species" and CN is a matrix-vector product. That is,

$$(CN)[i] = \sum_{j=1}^{\text{series}} C[i, j]N[j]$$

which allows us to identify the C matrix as the matrix of *competition coefficients*, with the requirement that $C[i,i] = 1$ (every species competes with itself with unit strength).

In the program "ecology" in Figure 9.8 we declare the following:

Species is the number of species in the model (default value = 11).
Community is a vector of population densities.
Threat is calculated as the term CN above, i.e., the total competition working against growth.
Velocity is the time derivative of the community vector.
Rate is the vector of rate constants r[i].
Capacity is the vector of carrying capacities k[i].
Competition is a matrix, with 1's on the main diagonal; generally c[i,j] is the competition coefficient for constraint of growth for species i caused by density of species j.

Figure 9.9 shows output for the program "ecology". You might have expected stability to be rare, causing the system to favor just one species. This is not the case. In fact, it is difficult to get even one extinction within the range of parameter values set up in the program "ecology", including the following ranges:

r[i] is set to 1, each i
k[i] is set to an input real cc, but the same real for each i
c[i,i] is set to 1 as it should be, but the other elements are randomly chosen in the range (0,0.3)

Figure 9.9 shows the output of the program "ecology", for input capacity 0.9, in the form of an 11×11 matrix followed by a printout of the numbers

community[k]   for k = 1 to 11

which are the densities for the ecology problem. It is evident that stability occurs for this case. Even this simple ecological program shows the phenomenon of stability arising from multiplicity of competitors.

BIOLOGICAL SIGNAL PROCESSING

In biological research it is often important to be able to process *digitized signals*. These are files of numbers that represent voltages, currents, etc., that arise directly or indirectly from living tissue. Examples of signals we may wish to process are as follows:

1. Biomedical signals, including electroencephalograph (EEG) and electro-cardiograph (EKG) voltages, and general organ-related signals
2. Neurological signals, which are similar to those in item 1 in that they arise from neurons, but are more fundamental (less organ specific)
3. Audio signals, including recordings of animal vocalizations
4. Signals emanating from primary machinery (spectrographs, preprocessors, cameras, etc.), which require secondary signal processing
5. Signals that are meant to trigger responses in living matter

Pascal programs cannot help us in *acquisition of data,* that is, in the creation of the number file representing the signal. Instead, we shall assume that such files exist (e.g., on discs) and that our task is to process the numbers. There are at least two ways to get the data into our programs. We can direct input to the program, that is, specify a file for input. This is usually accomplished in our operating systems. For a pure Pascal solution to the problem of getting the data from the file, we can write the statements

```
capacity constant: 0.9
competition matrix:
 1.000 0.250 0.088 0.155 0.212 0.238 0.265 0.219 0.141 0.215 0.046
 0.243 1.000 0.192 0.066 0.071 0.011 0.184 0.077 0.299 0.273 0.287
 0.163 0.138 1.000 0.079 0.191 0.143 0.024 0.250 0.073 0.290 0.188
 0.208 0.026 0.156 1.000 0.208 0.261 0.003 0.003 0.174 0.166 0.257
 0.126 0.155 0.084 0.122 1.000 0.014 0.014 0.129 0.202 0.123 0.141
 0.135 0.141 0.110 0.069 0.053 1.000 0.020 0.215 0.024 0.172 0.255
 0.214 0.283 0.107 0.259 0.257 0.100 1.000 0.242 0.016 0.097 0.059
 0.046 0.053 0.129 0.016 0.288 0.050 0.147 1.000 0.098 0.239 0.265
 0.130 0.220 0.200 0.030 0.179 0.175 0.144 0.182 1.000 0.009 0.273
 0.011 0.094 0.203 0.078 0.236 0.062 0.261 0.010 0.255 1.000 0.018
 0.132 0.086 0.182 0.227 0.262 0.240 0.067 0.033 0.253 0.107 1.000
```

time	s1	s2	s3	s4	s5	s6	s7	s8	s9	s10	s11
1.000	0.214	0.249	0.295	0.316	0.414	0.391	0.268	0.352	0.294	0.382	0.281
2.000	0.234	0.287	0.327	0.355	0.486	0.453	0.302	0.395	0.336	0.438	0.311
3.000	0.228	0.298	0.324	0.357	0.509	0.471	0.305	0.396	0.345	0.449	0.308
4.000	0.220	0.304	0.319	0.354	0.517	0.478	0.303	0.393	0.348	0.449	0.301
5.000	0.213	0.310	0.316	0.352	0.522	0.483	0.301	0.390	0.350	0.449	0.296
6.000	0.207	0.315	0.314	0.351	0.525	0.487	0.300	0.389	0.351	0.448	0.293
7.000	0.203	0.319	0.312	0.351	0.526	0.490	0.300	0.389	0.352	0.447	0.290
8.000	0.199	0.322	0.312	0.351	0.527	0.491	0.299	0.389	0.353	0.447	0.288
9.000	0.196	0.325	0.311	0.351	0.528	0.492	0.298	0.389	0.353	0.446	0.286
10.000	0.194	0.327	0.311	0.352	0.528	0.493	0.298	0.389	0.353	0.446	0.285
11.000	0.191	0.329	0.311	0.352	0.528	0.494	0.298	0.390	0.353	0.445	0.285
12.000	0.190	0.331	0.311	0.353	0.528	0.494	0.297	0.390	0.354	0.445	0.284
13.000	0.188	0.332	0.311	0.353	0.528	0.494	0.297	0.390	0.354	0.445	0.284
14.000	0.187	0.333	0.311	0.354	0.528	0.494	0.297	0.390	0.354	0.445	0.283
15.000	0.186	0.334	0.311	0.354	0.528	0.494	0.296	0.391	0.354	0.445	0.283
16.000	0.185	0.334	0.311	0.354	0.528	0.494	0.296	0.391	0.354	0.445	0.283
17.000	0.185	0.335	0.311	0.355	0.528	0.495	0.296	0.391	0.354	0.445	0.283
18.000	0.184	0.335	0.311	0.355	0.528	0.495	0.296	0.391	0.354	0.445	0.283
19.000	0.183	0.335	0.311	0.355	0.528	0.495	0.296	0.391	0.354	0.445	0.283
20.000	0.183	0.336	0.311	0.355	0.528	0.495	0.296	0.391	0.354	0.445	0.283
21.000	0.183	0.336	0.312	0.355	0.528	0.495	0.296	0.391	0.354	0.445	0.283
22.000	0.182	0.336	0.312	0.355	0.528	0.495	0.295	0.391	0.354	0.445	0.283
23.000	0.182	0.336	0.312	0.356	0.528	0.495	0.295	0.391	0.354	0.445	0.283

Figure 9.9

var f: text; (*var "f" will refer to the file containing signal*)
begin
      reset(f, 'bird data'); (*"f" is now a file of birdsong signal*)

The data can then be input using the statement

readln(f,x[n]); (*use 'readln(x[n])' instead for directed input*)

where the vars "x" and "n" must be properly declared and "n" must be incremented for each "readln".
    We next turn to the problem of declaring a signal. One way to do this is as follows:

const max = 1023; (*there will be 1024 possible data*)
type signal = array[0..max] of real; (*or "of integer"*)
var x: signal; n: integer;

With these declarations we can create a procedure similar to the one called "getvec" described in Appendix B or "getdata" described in Appendix C. The "getdata" procedure is more relevant, since we wish to set the array bound as part of the input procedure. A typical procedure might look like "getsig" in Figure 9.10. Simply calling the procedure

getsig(x,n);

will set up the signal array so that

$x[0], x[1], x[2], \ldots, x[n - 1]$

are the data and n is the total number of data even though the last index is $(n - 1)$.
    Next we will describe some manipulations of the signal data. The Fourier transform, as described in Chapter 6, is, for integer frequency indices k

$$q[k] = \sum_{j=0}^{N-1} x[j] \exp\left(\frac{2\pi ijk}{N}\right)$$

```
procedure getsig(var x: signal; var n: integer);
begin
n := 0;
repeat
     readln(f, x[n]);
     n := n + 1
until eof(f)
end; { getsig }
```

Figure 9.10

Note that as in Chapter 6 the sampling time has been absorbed into the definition of the signal array, that is, if the experimental real time between data is called tau then x[j] is the data taken at time j∗tau. As discussed in Chapter 6, the true frequency f corresponding to index k is f = k/(N∗tau). It turns out that the transform can be inverted, meaning that for given spectral numbers (generally complex)

$$q[0], q[1], q[2], \ldots, q[N-1]$$

we can always recover the z[j] by

$$x[j] = \frac{1}{N} \sum_{k=0}^{N-1} q[k] \exp\left(\frac{-2\pi ijk}{N}\right)$$

We usually do not take spectral data and try to recover the sequence x[j] with a program, but the inverse transform is indispensable in derivations of many useful formulas used in signal processing. The quantity called *autocorrelation,* for example, is easily analyzed using Fourier techniques. We shall investigate this procedure later.

Now we will do some simple processing with signal data. We start with the task of displaying the signal. The file

ekg.dat

is to contain the data for a simulated EKG voltage. The program "osc" in Figure 9.11 draws an *oscillogram* of the signal as it would appear on an oscilloscope. The output is to be read *lexicographically* (as you read a book), starting from the upper left. The plotter is made to denote the frames (of time), and we will record the time between samples as

tau = 0.030 second

meaning that the time between vertical bars in the output of the program "osc" is exactly 25.6 tau = 0.768 second. The program was run, and the internal Pascal statement

reset(f, 'ekg.dat');

allowed this particular file to be plotted as shown in Figure 9.12.

It is difficult to see certain details in the full oscillogram, which represents about 1 minute of heartbeat data. For this reason, we have a second program called "blowup" in Figure 9.13 that is designed to magnify two small adjacent segments of Figure 9.12. This program asks the question

frame (1-8) segment (0-9):

```
program osc(input, output);
(* oscillograph drawing routine, takes file of input data *)
const
    datscale = 30;   (* scale size of oscillograph excursions *)
    max = 2047;             (* sample size for oscillograph *)
type
    signal = array [0..max] of real;
var
    dat, t, dt, y: real;
    x: signal;
    f: text;
    n, row, k, nn: integer;

#include "plibh.i"

    procedure getsig(var x: signal; var n: integer);
    begin
    n := 0;
    repeat
        readln(f, x[n]);
        n := n + 1
    until eof(f)
    end; { getsig }

begin
    dt := 16 / max;            (* to fit the samples into the picture *)
    reset(f, 'ekg.dat');
    (* now do vertical strokes *)
    getsig(x, n);
    graph;
    for nn := 0 to 10 do begin
    plot(-1 + nn / 5, 1);
    draw(-1 + nn / 5, -0.8)
    end;
    nn := 0;                          (* this will be signal pointer *)
    for row := 1 to 8 do begin
    (* now set position of horizontal frame axis *)
    y := 1 - 0.25 * 1.8 / 2 * row + 0.12;
    t := -1;
    move(-1.3, y);
    alphal;
    writeln('frame ', row: 1);
    for k := 1 to trunc(max / 8) do begin
        dat := x[nn];
        nn := nn + 1;
        begin
            plot(t, y + dat / datscale)
        end;
        t := t + dt
    end
    end
end.
```

Figure 9.11

---

to which we answer

3   5

to get the segments 5 and 6 of frame 3. These are marked X on the full oscillogram (Figure 9.12). In the blowup in Figure 9.14 we can see the usual features of P, RS, and T waves (Stanford, 1975).

We will now do some real processing, and it will become clear why heartbeat signal has been chosen. There is a parameter—the beat or *pulse rate*—that we would like to extract with a Pascal program. We will need a frequency component in the signal of approximate value

$$f_p = \text{frequency} = \frac{1}{\text{period}} = \text{pulse rate} = \frac{1}{\text{time between pulses}}$$

One way to get this f_p is to perform a fast Fourier transform, as discussed in Chapter 6. To do this we run the program "fft" (Chapter 6, Figure 6.7) and use the parameters

2048 512 0.03 0 2

which are to be inserted as the first line of the data file "ekg.dat". If you are adept at handling input files with "reset" statements, you can modify the program "fft" in Chapter 6 to ask for the parameters as follows:

TOTAL # SAMPLES ? 2048
TOTAL SAMPLES PER SPECTRUM ? 512
TIME BETWEEN SAMPLES ? 0.03
FMIN, MINIMUM FREQUENCY ? 0
FMAX, MAXIMUM FREQUENCY ? 2

Of course, if you modify the program "fft" to provide a menu such as the one shown above, the data file "ekg.dat" should not be modified. The output for either

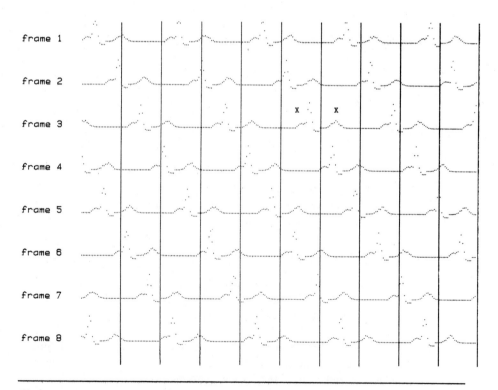

Figure 9.12 Simulated EKG data for 1 minute of real time, plotted by program "osc" (Figure 9.11) in oscillogram form. The segments marked X will be blown up by program "blowup" (Figure 9.13).

```
program blowup(input, output);
(* plot a 'blowup' of a large oscillogram ... the piece blown up
   will be 1/40 of the full gram *)
const
    max = 2047;                   (* all sizing keys to this max sample size *)
    datascale = 5;   (* make the data fit ! *)
type
    signal = array [0..max] of real;
var
    n, point, frame, segment: integer;
    seglength: integer;
    f: text;
    x: signal;

#include "plibh.i"

begin
    write('frame (1-8)    segment (0-9): ');
    (* compute # of samples per segment *)
    seglength := trunc(max / 80 + 0.5);
    readln(frame, segment);
    reset(f, 'ekg.dat');
    point := (frame - 1) * trunc(max / 8) + trunc(max / 80 * segment + 0.5);
    if point > 0 then
    for n := 0 to point - 1 do
        readln(f, x[0]);
    for n := 0 to 2 * seglength do
    readln(f, x[n]);
    graph;
    move(-1.3, 0.4);
    alphal;
    writeln('signal[n]');
    move(-1, 1);
    draw(-1, -1);
    move(-1, 0);
    draw(1, 0);
    for n := 0 to 2 * seglength do begin
    if n = 0 then
        move(-1, x[n] / datascale)
    else
        draw(-1 + n / seglength, x[n] / datascale)
    end
end.
```

Figure 9.13

method of entering the five parameters 2048, 512, 0.03, 0, 2 is the spectrum of the first 512 data of "ekg.dat" and appears as shown in the frequency-strength table in Figure 9.15. The output of the program "fft" tells us that

$f_p = 0.59$ beats/second

or so, corresponding to a pulse rate of nearly 36 pulses/minute. The other peaks in the spectrum are typical of real FFT data. There is usually some extra information, which is almost always one of the following types:

1. *Higher harmonics,* that is, multiple or approximate multiples of a fundamental. In the present case, the heartbeat signal has multiple pulses per period, which is what causes the strong harmonics.
2. *Aliasing,* meaning illegal peaks as a result of the finite sample size or, equivalently, of the finitude of the sampling time for FFT's. In general, you must respect the *Nyquist rule,* which states that frequency strengths for

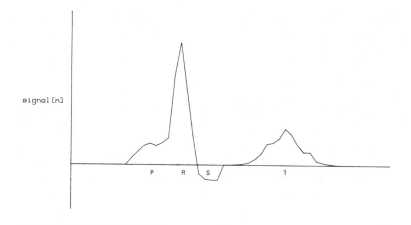

signal [n]

P   R   S   T

Figure 9.14   Frame 3, segments 5, 6 of the full oscillogram in Figure 9.12, is drawn here
by program "blowup" (Figure 9.13). The usual features of EKG signals are evident in this
simulation as signal transition regions P, RS, T. This figure illustrates one beat of the
heart.

```
Fast Fourier Transform output for first 512 data of EKG oscillogram
(ref. Chapter 6 for program "fft")

     freq.(Hz)      strength

       0.000       0.72008
       0.065       0.01751
       0.130       0.03250
       0.195       0.02684
       0.260       0.03581
       0.326       0.03359
       0.391       0.04173
       0.456       0.06416
       0.521       0.07758
       0.586       0.24565      -- this is pulse rate component, 35+ beats/min
       0.651       0.21051
       0.716       0.06835
       0.781       0.04269
       0.846       0.02814
       0.911       0.02441
       0.977       0.01442
       1.042       0.01246
       1.107       0.02066
       1.172       0.01190
       1.237       0.42142      -- this is second harmonic of pulse rate
       1.302       0.02998
       1.367       0.03540
       1.432       0.03258
       1.497       0.02959
       1.562       0.04820
       1.628       0.04951
       1.693       0.07412
       1.758       0.11035
       1.823       0.39268      -- this is third harmonic of pulse rate
       1.888       0.31084
       1.953       0.10543
```

Figure 9.15

components f having $f > (1/2)\text{tau}^{-1}$ are not legal. In other words, your data must sample at a rate that is at least twice as fast as the highest frequency you wish to analyze (Beauchamp, 1973).

It should be clear that the sampling rate $(\text{tau})^{-1}$ for the heartbeat oscillogram is about 33 hertz so we do not need to worry about aliasing in the FFT for the 0- to 2-hertz region. It is evident that the FFT is in agreement with the oscillogram itself in that the segments, as we have seen, are 0.768 second apart. Since the heartbeats come at just about every two segments, agreement is good.

The discussion of the heartbeat signal will be more complete if we include a computation of *autocorrelation*. The idea is to define autocorrelation coefficients

$$a[p] = \sum_{m=0}^{N-1} x[m]x[m+p]$$

We will invoke cyclic conditions and define $x[m+p]$ to wrap around to the start of the signal according to

$x[N]$ defined as $x[0]$
$x[N+1]$ defined as $x[1]$

and so on, with the general rule

$x[g] = x[g \bmod N]$

so that taking the index mod N always gives the cyclic conditions. It is interesting to note that if we now insert the inverse Fourier transform into the autocorrelation sum we find

$$a[p] = \frac{1}{N} \sum_{r=0}^{N-1} q[p]q^*[p] \exp\left(\frac{2\pi i r p}{N}\right)$$

This means that the autocorrelation is the Fourier transform of the signal consisting of relative powers at the various frequencies. That is, if the (artificial) signal

$$\frac{1}{N}|q[0]|^2, \frac{1}{N}|q[1]|^2, \ldots, \frac{1}{N}|q[N-1]|^2$$

were constructed, its Fourier spectrum would be the sequence of $a[p]$.

The importance of the $a[p]$ in signal processing is as follows. Note that the sum over $x[m]x[m+p]$ will be very large if the two x factors have, in each summand, the same sign. This will be the case for a periodic signal when p is the integer period. Take, for example, the signal defined by

$$x[m] = (-1)^m \qquad m = 0, \ldots, N-1 \text{ with N even}$$

```
program auto(input, output);
(* compute a table of autocorrelation coefficients a[p] from
    the signal data x[m] *)
const
    max = 2047;
type
    signal = array [0..max] of real;
var
    x: signal;
    n: integer;
    m, p: integer;
    sum: real;

    procedure getsig(var x: signal; var n: integer);
    begin
    n := 0;
    repeat
        readln(x[n]);
        n := n + 1
    until eof
    end; { getsig }

begin
    getsig(x, n);                (* fill up signal array with data and set n *)
    for p := 20 to 59 do begin
    sum := 0;
    for m := 0 to n - 1 do begin
        sum := sum + x[m] * x[(m + p) mod n]
    end;
    writeln(p: 4, sum: 4: 4)
    end
end.
```

Figure 9.16

so that the data are a sequence of +1's and −1's. Clearly, the autocorrelation coefficients a[0] and a[2] are large because terms separated by zero or two and, in fact, any even number of positions in the signal sequence have the same sign. However, a[1] = 0 and, in fact, all a[p] for odd p vanish.

We want to use similar ideas for the EKG signal in the file "ekg.dat". A program that computes autocorrelations is the program "auto" in Figure 9.16. We expect it to print out large values of a[p] for time lags p (actually, p*tau is the lag) that lie near the period between heartbeats. This output is, in fact, seen in Figure 9.17. The output clearly shows that the result for the heartbeat frequency is

$$f_p = \frac{1}{(tau*54)} = \frac{1}{1.62} = 0.617 \text{ hertz}$$

which is obtained from the high correlation for lag p = 54 shown in Figure 9.17. It is possible to see correlation coming from nearby peaks, but the only obvious and unobscured correlation is at lag 36, which is seen from the oscillogram or from the output of the program "blowup" (Figure 9.13) to be the distance from a T wave to the next R wave. Unfortunately, the tighter features of the EKG signal are swamped, as always happens, by the natural autocorrelation for small lags, which is always large. The obvious example is

a[0] = sum of all squared data values

```
Table of correlation coefficients for the EKG oscillogram

    lag p      coefficient

        20      299.5578     -- correlation between nearby peaks
        21      240.1761
        22      186.9792
        23      144.2267
        24      110.3182
        25       84.8158
        26       68.5670
        27       63.1312
        28       69.7572
        29       87.1969
        30      114.1375
        31      149.3929
        32      192.5203
        33      244.7240
        34      300.8107
        35      352.2878
        36      381.3230     -- correlation between T wave and NEXT R wave
        37      373.0958
        38      320.9552
        39      239.9375
        40      146.6272
        41       59.6719
        42       -6.7488
        43      -45.6694
        44      -53.0617
        45      -21.2611
        46       45.7931
        47      104.5034
        48      138.9189
        49      180.2324
        50      256.2974
        51      418.7671
        52      736.2543
        53     1137.9234
        54     1334.9201     -- this is the pulse rate correlation
        55     1132.0758        implying fp = 0.617 Hz. = 37.0 beats/min
        56      718.9358
        57      389.1338
        58      221.0813
        59      148.4874
```

Figure 9.17

which is essentially the power in the signal and must be the largest coefficient for any signal. There is no need to tabulate such values unless you want to know signal power. See the following exercises for more signal processing alternatives.

Exercises

1. Use the principle stated in the text that autocorrelation coefficients are the Fourier transform of the power spectrum (squares of direct Fourier coefficients) to generate the $a[p]$ by the following two-step process:
   (a) Perform an FFT on some initial signal data $x[m]$ to get $q[k]$.
   (b) Compute the sequence of squares $q[k]q^*[k]$, and perform another FFT on this new signal to get the autocorrelation $a[p]$.

   Compare the speed of this approach with the text approach (straightforward summation), especially for the largest signal index size N you can manage.

2. In the EKG oscillogram in the text the elapsed time for the whole plot is about 60 seconds. In principle, what is the greatest precision that this allows for the pulse rate f_p?

3. Obtain data for the rising and setting of the sun at your latitude over the course of 1 year. Use FFT or other techniques to evaluate the data. Find the mean length of a day, the length of a year (with trickery), and the amount of dithering in the length of a day, per day. A more complex example would be to evaluate data for high and low tides for the coast of Maine.

Answers

1. A good signal to try is

   ```
   x[n] := 0.5*gauss(0,1) + 0.5*x[n − 1]
   ```

   where x[0] is chosen to be zero and the rest of the data are generated recursively. The function gauss (,) is used as in the Appendix C library. The signal data are thus somewhat correlated. The sequence of steps a and b in exercise 1 should give the same autocorrelation as does the straightforward retardation sum in the text. The latter method is faster until the signal has a sufficiently large N.

2. In principle, the frequency precision is on the order of 1/60 hertz, i.e., the reciprocal of the measurement time. This means that spectral data for the problem will not vary appreciably over smaller frequency intervals; i.e., the frequency peaks are spread out by roughly this amount.

3. Refer to a good almanac for this data. There will be peaks twice per day for the tides, and these will advance a few minutes per day. For the rising and setting of the sun, your programs should be able to find the effect of the four seasons.

Appendixes

Appendixes

A | Graphics Libraries

OVERVIEW

The files suitable for graphics output are as follows:

plibg.i   Tektronix (Tek) 4012 terminal library
plibh.i   Hewlett-Packard (HP) 7470A plotter library
plib3.i   Three-dimensional library for either of the above devices
plibc.i   Cursor input library for the Tektronix 4012

All procedures are written in such a way that adaptation and modification for alternative output devices should be straightforward.

The ".i" name extensions are necessary for compilation on UNIX systems running Berkeley Pascal V1.2 but have no other significance.

General capabilities include ability to do the following:

1. *Clear* terminal screen (no effect for plotter)
2. Enter *graph* mode
3. Change back to *alpha*(numerics) mode
4. Draw *cir*(cles) of given radius and center
5. *Plot* a dot at x, y
6. *Draw* a line to x, y
7. *Move* invisible to x, y; as in preparing to draw
8. Draw *arrow*(s) and *arc*(s) with certain parameters

Additional capabilities of a device-dependent nature are described in what follows.

GRAPHICS SCREEN

All libraries use identical screen format for ease of programming and for portability. The coordinates are limited by

$$-1.3 \leq x \leq 1.3$$
$$-1 \quad \leq y \leq 1$$

and most of our applications simply bound all absolute coordinates with unity leaving the extra ± 0.3 for the x coordinate for alphanumerical labeling. The point (0,0) is always at screen center.

The manner in which we switch from graph to alpha mode depends on the output device, and hence the libraries differ somewhat in these procedures. Every effort has been made for compatibility among libraries. On the Reed College UNIX system, for example, we compile sources with the compile command

$ pas g source.p

for inclusion of "plibg.i",

$ pas h source.p

for inclusion of "plibh.i", and so on. Three-dimensional options are obtained by using

$ pas h 3 source

The important point is that this automatic inclusion of libraries is equivalent to the sequence

```
# include "plibh.i"
# include "plib3.i"
```

in the edited source *in the specified order.* Usually the compile or include option results in the same screen behavior for any of the output devices mentioned. With the aforementioned exception of alpha procedures, sources that run successfully on a Tektronix terminal are known to run equivalently on the HP plotter.

TEKTRONIX 4012 LIBRARY

This library is embodied in the include file

plibg.i

Specifications for the "plbg.i" procedures are as follows (see also Figure A.1):

Procedure	*Effect*
clear;	Erases the screen.
alpha;	Exits graphics mode to prepare labels.
graph;	Enters graphics mode.
draw(x,y: real);	Draws visibly to x, y. These coordinates are truncated as in the section titled *Graphics Screen* but are returned as they entered.
move(x,y: real);	Same as draw but invisible.
plot(x,y: real);	Puts a point at (x,y) with truncation as in the section titled *Graphics Screen.*

Procedure	*Effect*
arc(x,y,r,t1,t2: real);	Draws an arc clockwise from angle t1 to angle t2 centered at x, y with radius r.
arrow(x1,y1,x2,y2: real);	Draws an arrow pointing from x1, y1 to x2, y2.
cir(x,y,r: real);	Draws an arc of full 360 degree angle as in "arc".

```
(* plibg.i  :  Tektronix 4012 terminal library for graphics output *)
{ clear the Tek terminal screen }

    procedure clear;
    begin
        write(chr(27), chr(12))
    end; { clear }

{ put the Tek terminal in alpha mode }
    procedure alpha;
    begin
        write(chr(31))
    end; { alpha }

{ put the Tek terminal in graph mode }
    procedure graph;
    begin
        write(chr(29))
    end; { graph }

{ make a visible vector from last point to x,y }
    procedure draw(x, y: real);
    var
        hiy, loy, hix, lox: integer;
    begin
        if abs(x) > 1.3 then
            x := 1.3 * x / abs(x);
        if abs(y) > 1 then
            y := y / abs(y);
        x := 512 + 390 * x;
        y := 390 + 390 * y;
        hiy := trunc(y / 32);
        hix := trunc(x / 32);
        loy := 96 + trunc(y - 32 * hiy);
        hiy := hiy + 32;
        lox := 64 + trunc(x - 32 * hix);
        hix := hix + 32;
        write(chr(hiy), chr(loy), chr(hix), chr(lox))
    end; { draw }

{ make a hidden vector to x,y }
    procedure move(x, y: real);
    begin
        graph;
        draw(x, y)
    end; { move }
{ place a dot at x,y }

    procedure plot(x, y: real);
    begin
        move(x, y);
        draw(x, y)
    end; { plot }
```

Figure A.1

```
{ draw an arc centered at x,y with radius r, starting at angle tl
         and proceeding counter-clockwise to angle t2 }
    procedure arc(x, y, r, tl, t2: real);
    var
         t: real;
    begin
         t := tl;
         move(x + r * cos(tl), y + r * sin(tl));
         repeat
              t := t + 0.03141593;
              draw(x + r * cos(t), y + r * sin(t))
         until t >= t2
    end; { arc }
{ draw an arrow from x1,y1 to x2,y2 }

    procedure arrow(x1, y1, x2, y2: real);
    const
         PI = 3.1415926535;
    var
         theta, x, y: real;
    begin
         move(x1, y1);
         draw(x2, y2);
         x := x2 - x1;
         if x = 0 then begin
              theta := PI / 2;
              if y2 < y1 then
                   theta := -theta
         end else begin
              theta := arctan((y2 - y1) / x);
              if x < 0 then
                   theta := theta + PI
         end;
         x := x2 + 0.15 * sin(theta);
         y := y2 - 0.15 * cos(theta);
         arc(x, y, 0.15, theta + PI / 2, theta + 5 * PI / 8);
         x := x2 - 0.15 * sin(theta);
         y := y2 + 0.15 * cos(theta);
         arc(x, y, 0.15, theta - 5 * PI / 8, theta - PI / 2);
         move(x2, y2)
    end; { arrow }
{ draw a circle at x,y with radius r }

    procedure cir(x, y, r: real);
    var
         t: real;
    begin
         t := 0;
         move(x + r, y);
         repeat
              draw(x + r * cos(t), y + r * sin(t));
              t := t + 0.063
         until t > 6.6;
         move(x, y)
    end; { cir }
```

Figure A.1 (continued)

Any of the programs using these procedures will be equally successful with other output devices provided the calling syntax is preserved for the procedures, of which the first six are most common.

HEWLETT-PACKARD 7470A LIBRARY

This library is embodied in the include file

plibh.i

and has procedures as follows (see also Figure A.2.) (All x, y truncation follows the limits stated in the section titled *Graphics Screen.*)

Procedure	Effect
graph;	Sets up plotter so it has control over data lines. Procedure "alpha" below retrieves manual control, but labeling uses new procedure "alphal".
jump(x,y: real);	Moves pen to x, y but makes no change in pen status. This is internal and is not normally used in programs.
move(x,y: real);	A jump; pen is first lifted.
plot(x,y: real);	Puts a dot at x, y; pen stays down.
color(i: integer);	The i odd implies left pen, else right pen.
draw(x,y: real);	Draws to point x, y; pen stays down.
alphal;	Prepares next data string to be a label with a final return or "graph" call exiting back to graphics.
clear;	Moves pen out of the way for viewing.
cir; arrow; arc;	Identical to those for "plibg.i" in section titled *Tektronix 4012 Library.*
alpha;	Recovers manual terminal control; see "graph" above.
penvel(k: integer);	Pen velocity select; refer to listing.
linetype(c: integer);	Selects type of drawn lines; refer to listing and HP 7470A manual.

THREE-DIMENSIONAL LIBRARY

Each of the above graphics libraries or any compatible custom library that uses the same calling sequences can be made to plot three-dimensional graphics using the include file

plib3.i

which refers to the procedures

plot(x,y)
draw(x,y)
move(x,y)

but no others. Thus "plib3.i" must be included after "plibg.i", "plibh.i", or the user's compatible custom graphics inclusion.

```
procedure graph;

{ graph -- put HP plotter in programmed-on mode,  }
{ establish handshake protocol for communication }
{ of plotter instructions, and prepare to plot.  }
var
es: char;
begin
writeln;
es := chr(27);
write(es, '.Y');                   { Wake up plotter }
write(es, '.H40;5;6:');            { These device control commands }
write(es, '.M5;0;10;13:');         { establish format for   }
write(es, '.N2:');                 { plotter communication with programs. }

write('SC -13000,13000,-10000,10000;PU;');{ Set scale of P1,P2 and ...}
write('SP1;')                              { get the left pen. }
end;                                       { graph }

procedure jump(x, y: real);
{jump -- move pen to x,y without changing pen status. }

var
xx, yy: integer;
enq, ack: char;
begin
enq := chr(5);
if abs(x) > 1.3 then
    x := 1.3 * (x / abs(x));
if abs(y) > 1.0 then
    y := y / abs(y);

xx := trunc(10000 * x);        { Scale x and ... }
yy := trunc(10000 * y);        { scale y to fill plotting area. }
write(enq);                    { Check plotter buffer space }
readln(ack);                   { and wait until there is room for x,y.}
write('PA', xx: 1, ',', yy: 1, ';')     { Send the coordinates }
end;                                     { jump }

procedure move(x, y: real);

{ move -- pick up the pen and jump to x,y }
begin
write('PU;');
jump(x, y)
end;                                     { move }

procedure plot(x, y: real);

{ plot -- put a dot at x,y  }

begin
write('PU;');
jump(x, y);
write('PD;')
end;                                     { plot }
procedure color(i: integer);

{ color -- choose a pen; odd = left, even = right }

begin
if i mod 2 = 1 then
    write('SP1;')
else
    write('SP2;')
end;                                     {color}

procedure draw(x, y: real);

{ draw -- plot a line from current point to x,y }
begin
write('PD;');
jump(x, y)
end;                                     { draw }
```

Figure A.2

```
procedure alphal;

{ alphal -- label the plot with the next string output, }
{ until a RETURN is sent or the procedure graph is called. }

var
cr: char;
begin
cr := chr(13);
write('DI;');      { set label direction to default (horiz) status }
write('DT', cr, ';LB')           { enter label mode until RET is sent}
end;                                  {alphal}

procedure curse(j, i: integer);
{ curse -- cursor position for ISC programs }
begin
move(i * 2.6 / 79 - 1.3, j * 2.0 / 23 - 1.0)
end;                                  {curse}

procedure clear;
{ clear -- move the pen to the lower left corner of the plot. }
var
ec: char;
begin
ec := chr(27);
write(ec, '.Y');
write('PA250,279;');
write(ec, '.Z')
end;                                  { clear }

procedure cir(x, y, r: real);
{ cir -- draw a circle centered at x,y of radius r }

const
PI = 3.1415926;
var
ttp: real;
begin
move(x + r, y);
ttp := 0;
repeat
    draw(x + r * cos(ttp), y + r * sin(ttp));
    ttp := ttp + PI / 100
until ttp > 2 * PI;
move(x, y)
end;                                  { cir }

procedure arrow(x1, y1, x2, y2: real);
{ arrow -- draw an arrow from x1,y1 to x2,y2 }

var
run, rise: integer;
begin
move(x1, y1);
draw(x2, y2);
run := trunc(33 * (x2 - x1));
rise := trunc(30 * (y2 - y1));
write('DI',run,',',rise,';');{set direction for labeling arrow tip }
write('PU;');
write('CP-0.67,-0.25;');            { move to tip of line drawn }
writeln('LB>', chr(3))              { label the line with a '>' for a tip }
end;                                  { arrow }

procedure arc(x, y, r, t1, t2: real);
{ arc -- draw an arc centered at x,y with radius r and }
{ counter clockwise from angle t1 to angle t2      }
```

Figure A.2 (continued)

```
const
PI = 3.1415926;
var
ttp: real;
begin
ttp := t1;
move(x + r * cos(t1), y + r * sin(t1));
write('PD;');
while ttp <= t2 do begin
    ttp := ttp + PI / 100;
    jump(x + r * cos(ttp), y + r * sin(ttp))
end;
write('PU;')
end;                                          {arc}


{ The following procedures are unique to the HP plotter }

    procedure alpha;

    { alpha -- exit graphics mode, re-establish link }
    { between the terminal and unix                  }
    var
    ec: char;
    begin
    ec := chr(27);
    writeln;
    write('SP0;');                        { Put pen away }
    write(ec, '.Z')                       { Turn plotter off }
    end;                                  { alpha }


    procedure penvel(k: integer);
    { penvel -- set pen velocity to slower speed  }
    { for higher quality plotting.                }
    begin
    if k = 0 then
        write('VS;')          { reset velocity to default of 38.1 cm /sec }
    else if k < 127 then
        write('VS', k, ';')              { set velocity proportional to k }
    end;                                         { penvel }


    procedure linetype(c: integer);
    { linetype -- set line type }
    begin
    if (c > 0) or (c <= 6) then
        write('LT', c, ';');
    if c = 0 then
        write('LT;')
    end;                                          { linetype }
```

Figure A.2 (continued)

The library always plots the (x,y) component of the *rotated* vectors (x,y,z), where rotation is defined by *Euler angles:*

a   rotation around z axis
b   rotation around *new* x axis
c   rotation around *new* z axis

See Chapter 5, section titled *Three-Dimensional Graphics,* for examples of the use of the Euler angles. Note that procedure "rotate" will change the globals x, y, z but other procedures will not. (See Figure A.3.)

```
(* plib3.i    : 3-dimensional graphics library, refers back to
                plibg.i or plibh.i *)
{rotate causes vector reals x,y,z to rotate by angles
  a,b,c   around z,x,z axes resp.}

    procedure rotate(var x, y, z: real; a, b, c: real);
    var
        sx, sy, sz, tx, ty, tz: real;
    begin
        sx := x * cos(c) - y * sin(c);
        sy := x * sin(c) + y * cos(c);
        sz := z;
        tx := sx;
        ty := sy * cos(b) - sz * sin(b);
        tz := sy * sin(b) + sz * cos(b);
        x := tx * cos(a) - ty * sin(a);
        y := tx * sin(a) + ty * cos(a);
        z := tz;
    end; { rotate }

    procedure splot(x, y, z, a, b, c: real);
(* plot a point x,y,z after rotation by a,b,c *)
    var
        u, v, w: real;
    begin
        u := x;
        v := y;
        w := z;
        rotate(u, v, w, a, b, c);
        plot(u, v)
    end; { splot }

    procedure sdraw(x, y, z, a, b, c: real);
(* draw to x,y,z as it appears after rotation by a,b,c *)
    var
        u, v, w: real;
    begin
        u := x;
        v := y;
        w := z;
        rotate(u, v, w, a, b, c);
        draw(u, v)
    end; { sdraw }

    procedure smove(x, y, z, a, b, c: real);
(* move to point x,y,z as it appears after rotation by a,b,c *)
    var
        u, v, w: real;
    begin
        u := x;
        v := y;
        w := z;
        rotate(u, v, w, a, b, c);
        move(u, v)
    end; { smove }
    procedure axes(a, b, c: real);
(* draw the three coordinate axes in the orientation specified
  by a,b,c *)
    begin
        smove(-1, 0, 0, a, b, c);
        sdraw(1, 0, 0, a, b, c);
        smove(0, -1, 0, a, b, c);
        sdraw(0, 1, 0, a, b, c);
        smove(0, 0, -1, a, b, c);
        sdraw(0, 0, 1, a, b, c)
    end; { axes }
```

Figure A.3

Procedure	Effect
rotate(var x,y,z: real; a,b,c: real);	Rotates the vector x, y, z by a, b, c without plotting.
splot(x,y,z,a,b,c: real);	Plots the (x,y) part of the rotated vector x, y, z.
sdraw(x,y,z,a,b,c: real);	Draws to the (x,y) part.
smove(x,y,z,a,b,c: real);	Moves invisible to the (x,y) part.
axes(a,b,c: real);	Draws the x, y, z axes according to a, b, c.

The screen appearance for *null orientation* (a,b,c) = 0, 0, 0 is

+x to right
+y up
+z out toward user

The Euler angle "a" rotates counterclockwise, preserving the z axis absolutely, and so on. The matrix operations are

Rotate by a:
$$\begin{pmatrix} \cos(a) & -\sin(a) & 0 \\ \sin(a) & \cos(a) & 0 \\ 0 & 0 & 1 \end{pmatrix} \begin{pmatrix} x \\ y \\ z \end{pmatrix} = \begin{pmatrix} sx \\ sy \\ sz \end{pmatrix}$$

The coordinates sx, sy, sz are temporary local vars in "rotate".

Rotate by b:
$$\begin{pmatrix} 1 & 0 & 0 \\ 0 & \cos(b) & -\sin(b) \\ 0 & \sin(b) & \cos(b) \end{pmatrix} \begin{pmatrix} sx \\ sy \\ sz \end{pmatrix} = \begin{pmatrix} tx \\ ty \\ tz \end{pmatrix}$$

The tx, ty, tz are also local vars. The final transformation returns the final (x,y,z).

Rotate by c:
$$\begin{pmatrix} \cos(c) & -\sin(c) & 0 \\ \sin(c) & \cos(c) & 0 \\ 0 & 0 & 1 \end{pmatrix} \begin{pmatrix} tx \\ ty \\ tz \end{pmatrix} = \begin{pmatrix} x \\ y \\ z \end{pmatrix}$$

The actual plotting routines always display the (x,y) part of this last right-hand vector.

TEKTRONIX 4012 CURSOR INPUT

There is one procedure in the library "plibc.i" (see Figure A.4) called

curse(var x,y: real; var c: char);	Gets coordinates (x,y) and char c from crosshairs and keyboard, respectively.

When you call this procedure from within a program, the cursor display on the Tek 4012 appears. The procedure loops forever until a key is pressed, at which time the

```
(* plibc.i  :  Tektronix 4012 cursor input library *)
{ crosshair input procedure }
procedure curse (var x,y:real; var c:char);
(* strike a character on Tek keyboard and this procedure returns
   x,y as crosshair coordinates, char c as what you struck *)
var  absx,absy:integer;lox,loy,hix,hiy:char;
begin
        write(chr(27),chr(26));
        read(c,hix,lox,hiy);
        readln(loy);
        absx := (ord(lox) mod 32) + (32 *(ord(hix) mod 32));
        absy := (ord(loy) mod 32) + (32 *(ord(hiy) mod 32));
        x := (absx - 512)/390;
        y := (absy - 390)/390;
end;
```

Figure A.4

char "c" becomes the value of the key and the coordinates (x,y) are returned as numbers pertaining to crosshairs position. These numbers are precisely compatible with the graphics libraries' display procedures.

B | Matrix Library

SETTING UP THE TYPES

The most convenient way to begin a program that will use the "plibm.i" routines is

```
const dim  = 16; (*maximum dimension of matrices and vectors*)
type vector = array[1..dim] of real;
     matrix  = array[1..dim,1..dim] of real;
```

These types are sufficient for the library to run. Lack of dynamical arrays in Pascal forces us to use submatrices of matrix vars and subcolumns of vector vars. This truncation is done by the library routines.

Chapter 6 contains examples of the "plibm.i" routines, notably the "det" routine for determinants and the "changecol" column substitution routine for manipulation of linear equations.

LIBRARY FUNCTIONS

The library functions are as follows (see also Figure B.1):

Function	Effect
det(n: integer; m: matrix): real;	Returns the determinant of the (upper left-hand corner) n-by-n submatrix of m.
dot(n: integer; v1,v2: vector): real;	Returns the dot product of the vectors v1, v2; truncated after the first n component summands.
norm(n: integer; v: vector): real;	Returns sqrt(dot(n,v,v)).

Examples of the functions are as follows:
Let m be the matrix

$$m = \begin{pmatrix} 1 & 2 & 4 \\ 0 & 1 & 0 \\ 1 & 0 & 3 \end{pmatrix}$$

and v1, v2 be the vectors

$$v1 = \begin{pmatrix} 1 \\ 10 \\ 3 \end{pmatrix} \qquad v2 = \begin{pmatrix} 1 \\ 7 \\ 0 \end{pmatrix}$$

```
(* plibm.i  :  matrix library *)
(* requires globals type matrix = array[..,..] of real;
   vector = array[1..] of real *)

function det(n: integer; a: matrix): real;
var
        ii, jj, kk, ll, ff, nxt: integer;
        piv, cn, big, temp, term: real;
begin
        ff := 1;
        for ii := 1 to n - 1 do begin
        big := 0;
        for kk := ii to n do begin
                term := abs(a[kk, ii]);
                if term - big > 0 then begin
                big := term;
                ll := kk
                end
        end;
        if ii - ll <> 0 then
                ff := -ff;
        for jj := 1 to n + 1 do begin
                temp := a[ii, jj];
                a[ii, jj] := a[ll, jj];
                a[ll, jj] := temp
        end;
        piv := a[ii, ii];
        nxt := ii + 1;
        for jj := nxt to n do begin
                cn := a[jj, ii] / piv;
                for kk := ii to n + 1 do
                a[jj, kk] := a[jj, kk] - cn * a[ii, kk]
        end
        end;
        temp := 1;
        for ii := 1 to n do
        temp := temp * a[ii, ii];
        det := temp * ff
end; { det }

procedure minor(rows, columns, i, j: integer; a: matrix; var b: matrix);
(* put the matrix a, except for the i-th row and the j-th
column, into the matrix b.*)
var
        c1, c2, c3, c4: integer;
begin
        c3 := 0;
        for c1 := 1 to rows do
        if c1 <> i then begin
                c4 := 0;
                c3 := c3 + 1;
                for c2 := 1 to columns do
                if c2 <> j then begin
                        c4 := c4 + 1;
                        b[c3, c4] := a[c1, c2]
                end
        end
end; { minor }

procedure mmprod(n, m, k: integer; m1, m2: matrix; var m3: matrix);
(* multiply the n-by-m matrix in m1 by the m-by-k matrix in
m2, storing the result in m3.*)
var
        i, j, h: integer;
begin
        for i := 1 to n do
        for j := 1 to k do begin
                m3[i, j] := 0;
                for h := 1 to m do
                m3[i, j] := m3[i, j] + m1[i, h] * m2[h, j]
        end
end; { mprod }
```

Figure B.1

```
procedure smprod(rows, columns: integer; s: real; var v: matrix);
var
        i, j: integer;
begin
        for i := 1 to rows do
        for j := 1 to columns do
                v[i, j] := s * v[i, j]
end; { smprod }

procedure madd(rows, columns: integer; a, b: matrix; var c: matrix);
var
        i, j: integer;
begin
        for i := 1 to rows do
        for j := 1 to columns do
                c[i, j] := a[i, j] + b[i, j]
end; { madd }

procedure transpose(rows, columns: integer; var m: matrix);
var
        tmp: real;
        i, j: integer;
begin
        for i := 1 to rows do
        for j := i + 1 to columns do begin
                tmp := m[i, j];
                m[i, j] := m[j, i];
                m[j, i] := tmp
        end
end; { transpose }

procedure readmat(rows, columns: integer; var m: matrix);
var
        i, j: integer;
begin
        for i := 1 to rows do
        for j := 1 to columns do
                read(m[i, j])
end; { readmat }

procedure writemat(rows, columns: integer; m: matrix);
(* print the rows-by-columns matrix in m.*)
var
        i, j: integer;
begin
        for i := 1 to rows do begin
        for j := 1 to columns do
                write(m[i, j]);
        writeln
        end
end; { writemat }

function dot(n: integer; v1, v2: vector): real;
var
        i: integer;
        sum: real;
begin
        sum := 0;
        for i := 1 to n do
        sum := sum + v1[i] * v2[i];
        dot := sum
end; { dot }

function norm(n: integer; v: vector): real;
begin
        norm := sqrt(dot(n, v, v))
end; { norm }

procedure svprod(n: integer; s: real; var v: vector);
var
        i: integer;
begin
        for i := 1 to n do
        v[i] := v[i] * s
end; { svprod }
```

Figure B.1 (continued)

```
procedure vadd(n: integer; one, two: vector; var three: vector);
var
        i: integer;
begin
        for i := 1 to n do
        three[i] := one[i] + two[i]
end; { vadd }

procedure cross(one, two: vector; var three: vector);
{ the cross (or outer) product of the 3-vectors in 'one' and 'two'
                is placed in the vector 'three'.}
begin
        three[1] := one[2] * two[3] - one[3] * two[2];
        three[2] := one[3] * two[1] - one[1] * two[3];
        three[3] := one[1] * two[2] - one[2] * two[1]
end; { cross }

procedure readvec(n: integer; var v: vector);
var
        i: integer;
begin
        for i := 1 to n do
        read(v[i])
end; { readvec }

procedure writevec(n: integer; v: vector);
var
        i: integer;
begin
        for i := 1 to n do
        writeln(v[i])
end; { writevec }

procedure mvprod(m, n: integer; a: matrix; b: vector; var c: vector);
var
        i, j: integer;
        sum: real;
begin
        for i := 1 to m do begin
        sum := 0;
        for j := 1 to n do
                sum := sum + a[i, j] * b[j];
        c[i] := sum
        end
end; { mvprod }

procedure getrow(rows, columns, target: integer; m: matrix; var x: vector);
var
        i: integer;
begin
        for i := 1 to columns do
        x[i] := m[target, i]
end; { getrow }

procedure getcol(rows, columns, target: integer; m: matrix; var x: vector);
var
        i: integer;
begin
        for i := 1 to rows do
        x[i] := m[i, target]
end; { getcol }

procedure changerow(rows, columns, target: integer; var m: matrix; newrow: \
vector);
var
        i: integer;
begin
        for i := 1 to columns do
        m[target, i] := newrow[i]
end; { changerow }

procedure changecol(rows, columns, target: integer; var m: matrix; newcol: \
vector);
var
        i: integer;
```

Figure B.1 (continued)

```
begin
        for i := 1 to rows do
        m[i, target] := newcol[i]
end; { changecol }

procedure swaprow(rows, columns, row1, row2: integer; var m: matrix);
var
        i: integer;
        tmp: real;
begin
        for i := 1 to columns do begin
        tmp := m[row1, i];
        m[row1, i] := m[row2, i];
        m[row2, i] := tmp
        end
end; { swaprow }

procedure swapcol(rows, columns, col1, col2: integer; var m: matrix);
var
        i: integer;
        tmp: real;
begin
        for i := 1 to rows do begin
        tmp := m[i, col1];
        m[i, col1] := m[i, col2];
        m[i, col2] := tmp
        end
end; { swapcol }

procedure solve(n: integer; a: matrix; c: vector; var x: vector);
var
        k: integer;
        d: real;

        procedure swap(n, k: integer; var a: matrix; var c: vector);
        var
        e: real;
        j: integer;
        begin
        for j := 1 to n do begin
                e := c[j];
                c[j] := a[j, k];
                a[j, k] := e
        end
        end; { swap }

begin
        d := det(n, a);
        for k := 1 to n do begin
        swap(n, k, a, c);
        x[k] := det(n, a) / d;
        swap(n, k, a, c)
        end
end; { solve }

procedure invert(n: integer; a: matrix; var b: matrix);
(*invert the n-by-n matrix in a into b.*)
var
        I, tmp: vector;
        j: integer;
begin
        if det(n, a) = 0 then
        writeln('invert:singular matrix')
        else begin
        for j := 1 to n do begin
                I[j] := 1;
                solve(n, a, I, tmp);
                changecol(n, n, j, b, tmp);
                I[j] := 0
        end
        end
end; { invert }
```

Figure B.1 (continued)

Then

det(2,m)	returns 1
det(3,m)	returns −1
dot(2,v1,v2)	returns 71
dot(3,v1,v2)	returns 71
norm(3,v2)	returns sqrt(50)

LIBRARY PROCEDURES

There are many procedures in this library because of the large number of applications to which matrices present themselves. Cross product, for example, is done through a procedure because Pascal functions cannot return vector or matrix values. A list of procedures follows (see also Figure B.1):

Procedure	*Effect*
minor(rows, columns, i,j: integer; a: matrix; var b: matrix);	Forces matrix b to become a *minor* of matrix a, meaning the rows-by-columns part of a, except for the ith row and the jth column.
mmprod(n,m,k: integer; m1,m2: matrix; var m3: matrix);	General matrix multiply: m3 becomes the n-by-k matrix equal to the product of the n-by-m part of m1 and the m-by-k part of m2.
smprod(rows,columns: integer; s: real; var v: matrix);	Multiplies the matrix v by the *scalar* (real number) s and deposits the result only in the rows-by-columns part.
madd(rows,columns: integer; a,b: matrix; var c: matrix);	Matrix c becomes sum of matrices a and b but with only the rows-by-columns part involved in each matrix.
transpose(rows,columns: integer; var m: matrix);	*Transposes* the rows-by-columns part of matrix m. This is a flip of the main diagonal.
readmat(rows, columns: integer; var m: matrix);	Gets rows-by-columns part of matrix m by a series of readln calls.
writemat(rows,columns: integer; m: matrix);	Writes out the matrix m's rows-by-columns part using a series of write and writeln calls. The output format has a row on every teletype line.
vadd(n: integer; one, two: vector; var three: vector);	Vector "three" becomes the sum of the 1..n parts of vectors "one" and "two".
cross(one,two: vector; var three: vector);	Vector "three" becomes the *cross,* or *outer, product* of the 1..n parts of vectors "one" and "two".
readvec(n: integer; var v: vector);	Reads in the 1..n part of vector v.
writevec(n: integer; var v: vector);	Writes out the 1..n part of vector v.
mvprod(m,n: integer; a: matrix; b: vector; var c: vector);	Gets vector c's 1..m part as a result of m-by-n part of matrix a multiplying the 1..n part of vector b.

getrow(rows,columns,target: Forces the 1.. columns part of vector x to be the
integer; m: matrix; var x: vector);   target row of matrix m. Var "rows" is a
                           dummy.

getcol(rows, columns, target: Forces the 1..rows part of vector x to be the

Procedure	Effect
integer; m: matrix; var x:vector);	target column of matrix m. Var "columns" is a dummy.
changerow(rows, columns, target: integer; var m: matrix; newrow: vector);	Changes the target row of matrix m, in its 1..columns part, to agree with the 1..columns part of vector "newrow". Var "rows" is a dummy.
changecol(rows, columns, target: integer; var m: matrix; newrow: vector);	Changes the target column of matrix m, in its 1..rows part, to agree with the 1..rows part of vector "newrow". Var "rows" is a dummy.
swaprow(rows, columns, row1,row2: integer; var m: matrix);	Swaps the rows numbered "row1" and "row2" in matrix m, but only affects the 1..columns parts of each row. Var "rows" is a dummy.
swapcol(rows, columns, col1, col2: integer; var m: matrix);	Swaps the columns numbered "col1" and "col2" in matrix m, but only affects the 1..rows part of each column. Var "columns" is a dummy.
solve(n: integer; a: matrix; c: vector; var x: vector);	Solves the n equations in n unknowns $$a*x = c$$ where the x[j] are the unknowns, j = 1..n, and a, c are the coefficient matrix and constants vector, respectively.
invert(n: integer; a: matrix; var b: matrix);	Inverts a into b, meaning the n-by-n part. Error message if a is singular.

C | Statistics Library

GLOBAL VARS

The statistics library is embodied in the include file

plibs.i

and contains functions and procedures that act on "sample"s and their "size"s. The globals "sample" and "size" must be declared in order for the routines to work properly, for example,

```
type sample = array[1..1000] of real;
var x, y: sample;
    size: integer;
```

sets up "x" and "y" to be *statistical samples* that will ultimately be truncated by the variable "size". In this example declaration, the maximum size "1000" can be a Pascal constant such as "maxsize".

The functions and procedures of the library generally refer to the var "size" in much the same way that the matrix library (Appendix B) has internal reference to submatrices. In this way we bypass the lack of dynamical arrays in Pascal, as is explained in the following sections.

STATISTICAL FUNCTIONS

Library functions are as follows (see also Figure C.1):

Function	Effect
summ(var v: sample): real;	Returns the sum of all elements of sample v.
mean (var v: sample): real;	Returns summ/size, that is, the *mean* of sample v.
prod(var u,v: sample): real;	Returns the dot product of the samples u, v (as if they were vector types as in Appendix B).
error(var u: sample): real;	Standard deviation of the sample u truncated to "size" elements.
detr(var u: sample): real;	Returns sqr(summ(u)) − size*prod(u,u). This is an internal function used for linear regression.

(continued on page 218)

```
(* plibs.i  -  statistics library *)
(* requires globals "size: integer" and type sample = array[..] of real *)
(* arguments of type sample are declared var for efficiency reasons *)
    function summ(var v: sample): real;
    var
        ctr: integer;
        sum: real;
    begin
        sum := 0;
        for ctr := 1 to size do
            sum := sum + v[ctr];
        summ := sum
    end; { summ }

    function mean(var v: sample): real;
    begin
        mean := summ(v) / size
    end; { mean }

    function prod(var u, v: sample): real;
    var
        ctr: integer;
        sum: real;
    begin
        sum := 0;
        for ctr := 1 to size do
            sum := sum + u[ctr] * v[ctr];
        prod := sum
    end; { prod }

    function error(var u: sample): real;
    var
        m: real;
        ctr: integer;
        z: sample;
    begin
        m := mean(u);
        for ctr := 1 to size do
            z[ctr] := u[ctr] - m;
        error := sqrt(prod(z, z) / (size - 1))
    end; { error }

    function detr(var u: sample): real;
    begin
        detr := sqr(summ(u)) - size * prod(u, u)
    end; { detr }

    function bestm(var x, y: sample): real;
    var
        t: real;
    begin
        t := summ(y) * summ(x) - size * prod(x, y);
        bestm := t / detr(x)
    end; { bestm }

    function bestb(var x, y: sample): real;
    var
        t: real;
    begin
        t := summ(x) * prod(x, y) - summ(y) * prod(x, x);
        bestb := t / detr(x)
    end; { bestb }

    procedure getdata(var x: sample; var size: integer);
    begin
        size := 0;
        repeat
            size := size + 1;
            readln(x[size])
        until eof
    end; { getdata }
```

Figure C.1 (continued next page)

```
procedure getpairs(var x, y: sample; var size: integer);
begin
    size := 0;
    repeat
        size := size + 1;
        readln(x[size], y[size])
    until eof
end; { getpairs }

procedure putdata(var x: sample);
var
    ctr: integer;
begin
    for ctr := 1 to size do
        writeln(x[ctr])
end; { putdata }

procedure putpairs(var x, y: sample);
var
    ctr: integer;
begin
    for ctr := 1 to size do
        writeln(x[ctr], y[ctr])
end; { putpairs }

procedure trans(var x: sample; a, b: real);
var
    ctr: integer;
begin
    for ctr := 1 to size do
        x[ctr] := a * x[ctr] + b
end; { trans }

function maxpoint(var x: sample): integer;
var
    pt, ctr: integer;
    m: real;
begin
    m := x[1];
    pt := 1;
    for ctr := 2 to size do
        if x[ctr] > m then begin
            pt := ctr;
            m := x[ctr]
        end;
    maxpoint := pt
end; { maxpoint }

function minpoint(var x: sample): integer;
var
    ctr, pt: integer;
    m: real;
begin
    m := x[1];
    pt := 1;
    for ctr := 2 to size do
        if x[ctr] < m then begin
            pt := ctr;
            m := x[ctr]
        end;
    minpoint := pt
end; { minpoint }

procedure add(var x, y: sample);
var
    ctr: integer;
begin
    for ctr := 1 to size do
        x[ctr] := x[ctr] + y[ctr]
end; { add }
```

Figure C.1 (continued)

```
function poiss(mean: real): integer;
{produces random integers according to
the Poisson distribution.}
var
    sum: real;
    ctr: integer;
begin
    sum := 0;
    ctr := -1;
    repeat
        ctr := ctr + 1;
        sum := sum - ln(random(1))
    until sum > mean;
    poiss := ctr
end; { poiss }

function gauss(mean, variance: real): real;
var
    u, v, x: real;
begin
    repeat
        u := random(1);
        v := random(1);
        if u = 0.0 then
            u := 0.000000001;
        x := 2.0 * (v - 0.5) / u
    until sqr(x) <= -(4.0 * ln(u));
    gauss := x * sqrt(variance) + mean;
end; { gauss }
```

Figure C.1 (continued)

(continued from page 215)

Function	Effect
bestm(var x,y: sample): real;	Returns best-fit *slope* for the estimating equation y[n] = mx[n] + b, n = 1..size.
bestb(x,y: sample): real;	Returns best-fit *intercept* for the estimating equation y[n] = mx[n] + b, n = 1..size.
maxpoint(var x: sample): integer;	Returns *position*, not value, of greatest x[n] for the range n = 1..size.
minpoint (var x: sample): integer;	Returns *position*, not size, of smallest x[n] for the range n = 1..size.
poiss(mean: real): integer;	Returns *Poisson-distributed* random number; mean = "mean".
gauss(mean,variance: real): real;	Returns *gaussian-distributed* random number, mean = "mean" and error = square root of "variance".

STATISTICAL PROCEDURES

Many of the "plibs.i" procedures are designed for manipulation of samples in preparation for application of the above functions. Some of the procedures in the following table are used for input/output of samples (see also Figure C1).

Procedure	Effect
getdata(var x: sample; var size: integer);	Uses readln statements to get data into sample x. Var "size" is forced to be the length of the list x[1] x[2] x[3] . . . to be terminated with an end of file (EOF) character. Note the data format of one datum per line of teletype.
getpairs(var x,y: sample; var size: integer);	Arranges x, y to be filled with input data in the format x[1] y[1] x[2] y[2] x[3] y[3] . . . to be terminated with an EOF character. The var "size" is set in this case to be the number of lines of teletype.
putdata(var x: sample);	Outputs the data in the same format as described for procedure "getdata" above. The number of output data will be the current value of var "size".
putpairs (var x,y: sample);	Outputs data in the same format as described in "getpairs" above. Total number of pairs printed will be the value of var "size".
trans(var x: sample; a,b: real);	Replaces each element x[n] with the linear transformed value ax[n] + b; all of this for n = 1..size.
add(var x,y: sample);	Force sample x to be the elementwise sum x + y; that is, x[n] = x[n] + y[n] for n = 1..size.

USING THE FUNCTIONS

An example of a program that uses the standard statistical functions "mean" and "error" is

```
program easy(input,output);
const max = 100;
type sample = array[1..max] of real;
var x; sample; size: integer;
```

```
begin
      getdata(x,size);   (*gets elements of x, one per line on type, eventually
                            setting "size" to be the total number of data*)
      writeln('mean = p', mean(x));
      writeln('error = ',error(x));
end.
```

This program will respond to the input

1.3
2.5
0.951

(EOF character here, usually "ctrl-d" or "ctrl-c")

with the output of mean and error, as

mean = 1.5836666667e + 00
error = 8.1252712775e − 01

The functions "maxpoint" and "minpoint" are useful for *sorting*. A program segment that does a sort on a "sample" var is

```
(*x is a sample*)
temp := size; (*save "size" temporarily to prepare for sample truncation*)
while (size > 1) do begin
      j := maxpoint(x); (*an integer representing the index of greatest
                            x[n]*)
      if x[j] > x[size] then swap(j,size);
      size := size − 1;
end; size := temp; (*recover original value of "size"*)
```

The procedure "swap" is used as in Chapter 4, that is, it nondestructively interchanges the elements x[j] and x[size]. The above sort will arrange the sample x from smallest to largest elements, x[1] being smallest. The opposite sort will be done with "minpoint". It is important to remember that the actual maximum and minimum values to be found in a "sample" are

x[maxpoint(x)]   x[minpoint(x)]

respectively.
    The functions "poiss" and "gauss" are useful in a host of probability models, as described in Chapters 4 and 9. The Poisson distribution assumes that the probability of returning integer k is

$$P_k = \exp(-\text{mean}) \left(\frac{(\text{mean})^k}{k!} \right)$$

and it is easy to show that for the Poisson hypothesis thus quantified the overall mean is "mean." The Gaussian distribution is based on the *density*

$$p(x) = [2\pi \, (\text{variance})]^{-1/2} \exp \left(\frac{-(x - \text{mean})^2}{2*\text{variance}} \right)$$

with probability (x is in x, x + dx) equal to p(x) dx.

USING THE PROCEDURES

The following program segment assumes that x, y are of type "sample" as in the section titled *Global Vars,* with "size" declared integer:

```
getpairs(x,y,size);   (*this procedure sets "size"*)
putpairs(x,y);
```

This is an *echoing* program, which allows input such as

```
1    2
3.4    −8.9
6.8    −7.62
```

(terminate with EOF character, usually "ctrl-D" or "ctrl-C")

and then prints out the same data. The main point is that "getpairs" and "putpairs" are compatible, with the former setting "size" automatically, as it accepts pairs in the format

```
x[1]    y[1]
x[2]    y[2]
...
```

Procedure "getdata", which is even simpler, is used to set just one "sample".

After you obtain full "samples" using such procedures, the functions such as "mean(x)", "error(x)", "bestm(x,y)", and "bestb(x,y)" will produce statistical results. For example, right after the "putpairs" procedure above, or in place of it, we can write the statement

```
writeln(bestm(x,y),bestb(x,y));
```

to find out the best m and b for the approximation y[n] = mx[n] + b, with n going from 1 to size; in the above case n goes from 1 to 3.

One practical use of the procedures and functions combined is to graph arbitrary pairs of samples x, y so that all data are within graphics screen bounds. The "trans" procedure is ideal for scaling an entire sample, as follows. First, note that x[maxpoint(x)] and x[minpoint(x)] are extreme data values, and −1 to +1 is the screen range we usually desire. Therefore, in order to fit an entire sample on the screen, we write

```
trans(x,1/xscale,xoff);
```

where we define

$$\text{xscale} = (1/2)(x[\text{maxpoint}(x)] - x[\text{minpoint}(x)])$$

$$\text{xoff} \;\;\; = -1 - \frac{(x[\text{minpoint}(x)])}{\text{xscale}}$$

This works in this case because the procedure "trans" warps the sample x in such a way that the least value x[n] becomes -1 and the greatest value x[n] becomes $+1$. Thus the linear transformation described is the unique one that exactly fills the screen.

D | Special Functions Library

CONTENTS

The special functions library is embodied in the include file

plibl.i

and has no procedures, only functions (see Figure D.1). The definitions of all functions of the library follow Chapter 2 reference Abramowitz and Stegun (1965) and are virtually compatible with Chapter 2 reference Gradshteyn and Ryzhik (1965).

GAMMA FUNCTION

The *gamma function* is so fundamental that it is first in the library. It is referenced by several other functions such as the Bessel functions that follow it. The calling syntax is

gam(x: real): real;   for $\Gamma(x)$

The method of computation is power series for $1/\Gamma(x)$, which is very accurate for $0 < x < 1/2$. For other values of x, either the relation

$$\Gamma(n + 1) = n\Gamma(n)$$

or one of the *reflection formulas* of the references is used to reduce the problem to the interval $(0,1/2)$. The function is *recursive*; a unique example of such functions in that the gamma function has unique interrelations. The function is valid for all real arguments not equal to integers $0, -1, -2, -3, \ldots$.

BESSEL FUNCTION

The calling for the Bessel function is

j(nu z: real): real;   for $J_{nu}(z)$

```
(* plibl.i   - special functions library *)

    function gam(x: real): real;
(* THE STANDARD GAMMA FUNCTION *)
    const
        pi = 3.1415926535897932;
    var
        y, f: real;

        function sgam(y: real): real;
        var
            tmp, f: real;
        begin
            f := y;
            tmp := f;
            f := f * y;
            tmp := tmp + f * 0.5772156649015329;
            f := f * y;
            tmp := tmp + f * (-0.6558780715202538);
            f := f * y;
            tmp := tmp + f * (-0.0420026350340952);
            f := f * y;
            tmp := tmp + f * 0.1665386113822915;
            f := f * y;
            tmp := tmp + f * (-0.0421977345555443);
            f := f * y;
            tmp := tmp + f * (-0.009621971527877);
            f := f * y;
            tmp := tmp + f * 0.0072189432466630;
            f := f * y;
            tmp := tmp + f * (-0.0011651675918591);
            f := f * y;
            tmp := tmp + f * (-0.0002152416741149);
            f := f * y;
            tmp := tmp + f * 0.0001280502823882;
            f := f * y;
            tmp := tmp + f * (-0.0000201348547807);
            f := f * y;
            tmp := tmp + f * (-0.0000012504934821);
            f := f * y;
            tmp := tmp + f * 0.0000011330272320;
            f := f * y;
            tmp := tmp + f * (-0.0000002056338417);
            f := f * y;
            tmp := tmp + f * 0.0000000061160950;
            f := f * y;
            tmp := tmp + f * 0.0000000050020075;
            f := f * y;
            tmp := tmp + f * (-0.0000000011812746);
            f := f * y;
            tmp := tmp + f * 0.0000000001043427;
            f := f * y;
            tmp := tmp + f * 0.0000000000077823;
            f := f * y;
            tmp := tmp + f * (-0.0000000000036968);
            f := f * y;
            tmp := tmp + f * 0.00000000000005100;
            sgam := 1 / tmp
        end; { sgam }

        begin
            if x = 1 then
                gam := 1
            else begin
                if x < 0 then
                    gam := -(pi / (x * sin(pi * x) * gam(-x)))
                else begin
                    y := x;
                    f := 1;
                    while y > 1 do begin
                        y := y - 1;
                        f := f * y
                    end;
```

Figure D.1

```
                    if y = 1 then
                        gam := f
                    else begin
                        if y > 1 / 2 then
                            gam := f * pi / (sin(pi * y) * gam(1 - y))
                        else
                            gam := f * sgam(y)
                    end
                end
            end
    end; { gam }

    function j(nu, z: real): real;
(* THE BESSEL FUNCTION OF ORDER nu *)
    var
        n, d, f, sum: real;
        ctr: integer;
    begin
        if z = 0 then begin
            if nu = 0 then
                j := 1
            else
                j := 0
        end else begin
            f := 1 / gam(nu + 1);
            n := -(sqr(z) / 4);
            d := nu + 1;
            sum := f;
            ctr := 1;
            repeat
                f := f * n / (ctr * d);
                d := d + 1;
                ctr := ctr + 1;
                sum := sum + f
            until abs(f) < 0.0000000000001;
            if z > 0 then
                j := sum * exp(nu * ln(z / 2))
            else
                j := (1 - 2 * (trunc(nu) mod 2)) * sum *
                    exp(nu * ln(abs(z / 2)))
        end
end; { j }

    function tan(x: real): real;
    begin
        tan := sin(x) / cos(x)
    end; { tan }

    function cosh(z: real): real;
    begin
        cosh := (exp(z) + exp(-z)) / 2
    end; { cosh }

    function sinh(z: real): real;
    begin
        sinh := (exp(z) - exp(-z)) / 2
    end; { sinh }

    function sech(z: real): real;
    begin
        sech := 1 / cosh(z)
    end; { sech }

    function arctanh(x: real): real;
    begin
        arctanh := 0.5 * ln((1 + x) / (1 - x))
    end; { arctanh }

    function erf(x: real): real;
    var
        temp, mult, t, a1, a2, a3, a4, a5, p: real;
```

Figure D.1 (continued)

```
      begin
          if x < 0 then
              mult := -1
          else
              mult := 1;
          x := abs(x);
          p := 0.3275911;
          a1 := 0.254829592;
          a2 := -0.284496736;
          a3 := 1.421413741;
          a4 := -1.453152027;
          a5 := 1.061405429;
          t := 1 / (1 + p * x);
          temp := a1 * t + a2 * t * t + a3 * t * t * t +
                  a4 * t * t * t * t + a5 * t * t * t * t * t;
          erf := mult * (1 - exp(-sqr(x)) * temp)
      end; { erf }

      function erfr(x, y: real): real;
(* THE REAL PART OF COMPLEX ERROR FUNCTION , z = x + iy *)
      const
          pi = 3.14159265359;
      var
          sum, piece, f: real;
          n: integer;
      begin
          sum := erf(x);
          if abs(x) < 0.0000001 then
              piece := 0
          else begin
              piece := exp(-sqr(x)) / x * (1 / (2 * pi));
              piece := piece * (1 - cos(2 * x * y))
          end;
          sum := sum + piece;
          piece := 0;
          for n := 1 to 10 do begin
              f := 2 * x - 2 * x * cos(2 * x * y) * cosh(n * y) +
                  n * sinh(n * y) * sin(2 * x * y);
              piece := piece + exp(-(sqr(n) / 4)) / (sqr(n) + 4 * sqr(x)) * f
          end;
          sum := sum + piece * 2 / pi * exp(-sqr(x));
          erfr := sum
      end; { erfr }

      function erfi(x, y: real): real;
(* THE IMAGINARY PART OF COMPLEX ERROR FUNCTION, z = x + iy *)
      const
          pi = 3.14159265359;
      var
          sum, piece, g: real;
          n: integer;
      begin
          sum := 0;
          if abs(x) < 0.0000001 then
              piece := y / pi
          else begin
              piece := exp(-sqr(x)) / (2 * pi * x) * sin(2 * x * y)
          end;
          sum := sum + piece;
          piece := 0;
          for n := 1 to 20 do begin
              g := 2 * x * cosh(n * y) * sin(2 * x * y) +
                  n * sinh(n * y) * cos(2 * x * y);
              piece := piece + exp(-(sqr(n) / 4)) / (sqr(n) + sqr(2 * x)) * g
          end;
          sum := sum + piece * exp(-sqr(x)) * 2 / pi;
          erfi := sum
      end; { erfi }
```

Figure D.1 (continued)

```pascal
    function f(a, b, c, z: real): real;
(* GAUSS HYPERGEOMETRIC FUNCTION *)
    var
        g, sum: real;
        n: integer;
    begin
        g := 1;
        n := 1;
        sum := 0;
        repeat
            sum := sum + g;
            g := g * a * b * z / (n * c);
            n := n + 1;
            a := a + 1;
            b := b + 1;
            c := c + 1
        until (n > 40) or (abs(g) < 1e-10);
        f := sum
end; { f }

    function m(a, b, z: real): real;
(* KUMMER (CONFLUENT) HYPERGEOMETRIC FUNCTION *)
    var
        g, sum: real;
        n: integer;
    begin
        sum := 0;
        g := 1;
        n := 1;
        repeat
            sum := sum + g;
            g := g * a * z / (b * n);
            n := n + 1;
            a := a + 1;
            b := b + 1
        until (n > 40) or (abs(g) < 1e-10);
        m := sum
    end; { m }

    function u(a, b, z: real): real;
(* ASSOCIATED CONFLUENT HYPERGEOMETRIC FUNCTION *)
    const
        pi = 3.1415926535897932;
    begin
        if abs(z) < 1e-10 then
            u := 0
        else if (abs(b - trunc(b)) < 0.0001)
                or (abs(b - round(b)) < 0.0001) then
                u := (u(a, b - 0.0005, z) + u(a, b + 0.0005, z)) / 2
        else
            u := pi / sin(pi * b) *
                (m(a, b, z) / (gam(1 + a - b) * gam(b)) -
                exp((1 - b) * ln(z)) *
                m(1 + a - b, 2 - b, z) / (gam(a) * gam(2 - b)))
    end; { u }

    function her(n: integer; x: real): real;
(* HERMITE POLYNOMIAL OF ORDER n *)
    begin
        if n = 1 then
            her := 2 * x
        else
            her := x * exp(n * ln(2)) * u(0.5 - n / 2, 3 / 2, sqr(x))
    end; { her }

    function leg(n: integer; x: real): real;
(* LEGENDRE POLYNOMIAL OF ORDER n *)
    begin
        leg := f(-n, n + 1, 1, (1 - x) / 2)
    end; { leg }
```

Figure D.1 (continued)

```
    function lag(n: integer; a, x: real): real;
(* LAGUERRE FUNCTION OF ORDER n, SUPERSCRIPT a *)
    var
        fact: real;
    begin
        if n mod 2 = 0 then
            fact := 1
        else
            fact := -1;
        lag := u(-n, a + 1, x) / gam(n + 1) * fact
    end; { lag }

    function i(nu, x: real): real;
(* MODIFIED BESSEL FUNCTION I OF ORDER nu *)
    begin
        i := exp(-x) * exp(nu * ln(x / 2)) *
            m(nu + 0.5, 2 * nu + 1, 2 * x) / gam(nu + 1)
    end; { i }

    function k(nu, x: real): real;
(* MODIFIED BESSEL FUNCTION K OF ORDER nu *)
(* positive args only *)
    const
        pi = 3.1415926535897932;
    begin
        k := exp(nu * ln(2 * x)) * exp(-x) * sqrt(pi) *
            u(nu + 0.5, 2 * nu + 1, 2 * x)
    end; { k }

    function comb(n: real; m: integer): real;
(* COMBINATORIAL BRACKET n-over-m, m AN INTEGER *)
    begin
        if m = 1 then
            comb := n
        else
            comb := comb(n - 1, m - 1) * n / m
    end; { comb }

    function jac(n: integer; a, b, x: real): real;
(* JACOBI POLYNOMIAL OF ORDER n, SUPERSCRIPTS (a,b) *)
    begin
        jac := comb(n + a, n) * f(-n, n + a + b + 1, a + 1, (1 - x) / 2)
    end; { jac }

    function tcheb(n: integer; x: real): real;
(* TCHEBYSHEV POLYNOMIAL OF ORDER n *)
    begin
        tcheb := f(-n, n, 0.5, (1 - x) / 2)
    end; { tcheb }
```

Figure D.1 (continued)

The computation is done by power series. The ranges of argument and order are as follows:

Order nu	Argument z
Any integer	Any real number
Noninteger	Any positive real

In the illegal cases, the calculation is not divergent, just erroneous.

SIMPLE TRANSCENDENTAL FUNCTIONS

The functions

```
tan(x: real): real;        for tan(x)
cosh(z: real): real;       for cosh(z)
sinh(z: real): real;       for sinh(z)
sech(z: real): real;       for sech(z)
arctanh(x: real): real;    for arctanh(x)
```

have been included in "plibl.i" for convenience or for reference by other library functions.

ERROR FUNCTIONS

The two functions

```
erfr(x,y: real): real;   (*real part*)       for erf(x + iy)
erfi(x,y: real): real;   (*imaginary part*)
```

are calculated by a hyperbolic-trigonometric series given in Chapter 2 reference Abramowitz and Stegun (1965).

GAUSS HYPERGEOMETRIC FUNCTION

The function

$$F(a,b,c,z) = 1 + \frac{abz}{c} + \frac{a(a+1)b(b+1)z^2}{c(c+1)2!} + \cdots$$

is especially useful for calculating other functions, especially those of mathematical physics. In "plibl.i" the function is denoted

```
f(a,b,c,z: real): real;    for F(a,b,c,z)
```

Various values of arguments rule out the possibility of convergence. In particular, c should not be a negative integer. Chapter 2 reference Abramowitz and Stegun (1965) has exact convergence criteria. The astute programmer will be able to enhance convergence by using the multitude of interrelations and variable tranformations enjoyed by the gaussian function.

KUMMER HYPERGEOMETRIC FUNCTION

The Kummer function

$$M(a,b,z) = 1 + \frac{az}{b} + \frac{a(a+1)}{b(b+1)} \frac{z^2}{2!} + \cdots$$

and its associate

$$U(a,b,z) = \frac{\pi}{\sin(\pi b)} \left(\frac{M(a,b,z)}{\Gamma(1+a-b)\Gamma(b)} - z^{1-b} \frac{M(1+a-b, 2-b, z)}{\Gamma(a)\Gamma(2-b)} \right)$$

are, like the gaussian function F, useful for computation of other functions. The Kummer functions are especially applicable in computations of *orthogonal polynomials*. In "plibl.i" the Kummer functions are computed with the same loop criteria as is F. It is important that z be positive in the call to the U function. This is handled properly by the orthogonal polynomials blocks in the remainder of "plibl.i". The calling for the Kummer functions is

```
m(a,b,z: real): real;     for M(a,b,z)
u(a,b,z: real): real;     for U(a,b,z)
```

ORTHOGONAL POLYNOMIALS

The orthogonal polynomials of

Hermite	her(n: integer; x: real): real;	for $H_n(x)$
Legendre	leg(n: integer; x: real): real;	for $P_n(x)$
Laguerre	lag(n: integer; a,x: real): real;	for $L_n^a(x)$
Jacobi	jac(n: integer; a,b,x: real): real;	for $P_n^{ab}(x)$
Tchebyshev	tcheb(n: integer; x: real): real;	for $T_n(x)$

are calculated in terms of hypergeometric functions in "plibl.i".

Precision is exact in the sense that the series always terminate for the given polynomials. If you desire the *integer* coefficients of these orthogonal polynomials, you can solve determinant equations (using, e.g., the library in Appendix B) in order to get near-integer coefficients n', then do n = trunc(n' + 0.01) or the like to fix integer values even though the functions all return reals. Alternatively, you can use recurrence relations or Rodriguez-type formulas for generating coefficients.

These polynomials provide good checking functions for integration techniques since they are orthonormal with appropriate weight functions in their respective inner products.

MODIFIED BESSEL FUNCTIONS

Modified Bessel functions are

i(nu,x: real): real;   for $I_{nu}(x)$
k(nu,x: real): real;   for $K_{nu}(x)$

These are likewise computed from hypergeometric series. Only positive real arguments are allowed.

COMBINATORIAL BRACKET

The bracket

$$\binom{n}{m} = \frac{n!}{m!(n-m)!}$$

is computed recursively for any integer m with $n \geq m$. The call is

comb(n: real; m: integer): real;   for $\binom{n}{m}$

Note that n can be nonintegral, in which case the bracket is interpreted as the product of exactly m terms of the form

$$\frac{n-j+1}{m-j+1}$$

as $j = 1, \ldots, m$.

E | Dynamical Models Library

LIBRARY CAPABILITIES

The library embodied in the include file

plibd.i

was designed to aid in the modeling of dynamical phenomena. There are two main capabilities:

1. To model real *time-evolving functions* f(x,t) by plotting three-dimensional views of the x, t, and f axes. Examples of such functions are components of wave equations, components of diffusion equations, and population functions. Often the function f is obtained numerically from a differential equation.
2. To model *ensembles* of coordinates by calculating the change in position and velocity of each of many particles given their respective accelerations. Examples of the use of such routines are classical mechanics modeling, N-body problems, and dynamical vector problems in general.

REQUIRED GLOBALS

For capability 1 above, the required globals are typically declared as

```
const max = 100;    (*the time-evolving function will be f[-100..100]*)
type wave = array[-max..max] of real;
var f :wave;             (*f will be the plotted time-evolving function*)
```

Since the one space dimension x and the time axis t are also to be drawn, we usually declare "x, t: real" as well.

For capability 2 above, the required globals are typically declared as

```
const dt = 0.0001;     (*the time increment for equation solving*)
      dim = 3;          (*choice of three space dimensions*)
      num = 25;         (*choice of 25 particles per ensemble*)
type ensemble = array[1..num,1..dim] of real;
var position, velocity, acceleration: ensemble;
```

This last declaration sets up the relevant mechanical data for each of 25 particles.

PROCEDURES

There are three procedures of the library; the first two are involved in time-evolution plots (capability 1) and the third addresses capability 2:

Procedure	Effect
ax;	Draws the x, t, and f axes to prepare for plots of f(x,t). The view orientation is determined by Euler angles a, b, c, but the angles have been fixed empirically for optimal viewing.
frame(f: wave; t: real);	Draws a single "frame" defined as the set of array values f[−max..max] calculated with respect to the time variable t.
advance(var e: ensemble; de: ensemble);	Computes componentwise the formal time advance of the ensemble e, for example, as
	e := e + de*dt;
	Updates all positions according to present velocities.

TIME EVOLUTION PLOTS

Let the procedures "ax" and "frame" be considered as a sublibrary that is called as follows:

plibdl.i   (*number '1' refers to capability 1*)

We call it in this way so that "ensembles" need not be defined. The procedures now allow general *space-time plots*, that is, drawings of functions that satisfy space-time differential equations.

The program "spacetime" in Figure E.1 shows a useful skeleton which we can use for Pascal statements for particular problems. The only additions required to program "spacetime" are

1. Insertion of an initialization loop inside block of procedure "setup"
2. Insertion of an update segment in main block's "repeat-until" loop

Note that certain constants have been worked out for optimal viewing of the space-time plots. These predetermined constants are

```
program spacetime(input,output);
(* skeleton program for space-time evolution analyses *)
const max=100;
      dt=0.1;
type wave=array[-max..max] of real;
var psi:wave;
    t:real;

# include "plibh.i"  (* include plotter library *)
# include "plib3.i"  (* include 3-dimensional library *)
# include "plibdl.i" (* include dynamical proc's 'ax' and 'frame' only *)
procedure setup(var psi:wave);
    var ii:integer;
    begin
            for ii:= -max to max do
                      (* setup psi with first picture *)
    end;
begin
    graph;
    ax;
    setup(psi);
    frame(psi,0);
    repeat
    t:=t+dt;
    (* go and update array psi right here *)
    frame(psi,t);
    until t>1.9
end.
```

Figure E.1

1.9	(undeclared)	The limit on t for good plots
0	(undeclared)	The starting value of t recommended
0.85, 0.8, −1		Euler angles "a,b,c" for viewing

The procedure "frame" in "plibd.i" (and "plibd1.i") takes care of the plotting of function values so that in program "spacetime"

psi = 0	Plotted on the (x,t) plane
psi = 1	Plotted near top of the paper
psi = −1	Plotted near bottom of the paper

The program "diffuse" in Figure E.2 uses all of these considerations in plotting an exact solution of the diffusion equation (Figure E.3)

$$\frac{\partial(\text{psi})}{\partial t} = D \frac{\partial^2}{\partial x^2} (\text{psi})$$

for *diffusion constant D*. The solution plotted is

$$\text{psi}(x,t) = (4\pi Dt)^{-1/2} \exp\left(\frac{-x^2}{4Dt}\right)$$

```
program diffuse(input, output);
(* demonstration program    -    one-dimensional diffusion *)
(* "psi" is the density, and diffusion is governed by:

          (d/dt) psi = D (d/dx)(d/dx) psi ,

   where D is diffusion constant *)
(* the differential equation has been presolved - this program
   just plots the solution *)
const
      max = 100;                         (* psi will have indices from -max to +max *)
      D = 2.5;                       (* Einstein diffusion constant *)
      aa = -1;
      bb = 0.8;
      cc = 0.85;                    (* these are needed only for labels *)
      dt = 0.1;
type
      wave = array [-max..max] of real;
var
      psi: wave;
      ii: integer;
      t: real;
#include "plibh.i"
#include "plib3.i"
#include "plibdl.i"

      procedure setup(var psi: wave);
      var
       ii: integer;
      begin
       for ii := -max to max do
           psi[ii] := 0.7 * exp(-(sqr(ii) * 40 / sqr(max)));
      end; { setup }

begin
      graph;
      ax;
      smove(1, 0.46, 0, aa, bb, cc);
      alphal;
      writeln('X AXIS');
      smove(0.06, 1, 0, aa, bb, cc);
      alphal;
      writeln('PSI AXIS');
      smove(-0.38, 0.26, 1, aa, bb, cc);
      alphal;
      writeln('T AXIS');
      setup(psi);
      frame(psi, 0);
      repeat
       t := t + dt;
       (* now we can update psi for the new time t *)
       for ii := -max to max do
           psi[ii] := 1.4 / sqrt(4 + 4 * D * t) *
                        exp(-(sqr(ii) * 40 / (sqr(max) * (1 + D * t))));
       frame(psi, t);
      until t > 1.9
end.
```

Figure E.2

except that it is normalized (multiplied by 0.7) and time is started away from the singular temporal origin in the program "diffuse". Note that in program "diffuse" there are labels placed on the picture (Figure E.2. The "smove" procedures required to position the labels have arguments derived from the special values of a, b, c given above for axes lines.

Note that the solution psi(x,t) is easily calculated, and often this sort of foreknowledge of the wave function is absent in a problem. Instead, we often have to solve a differential equation, which is why the extra procedure "advance" was placed in the library.

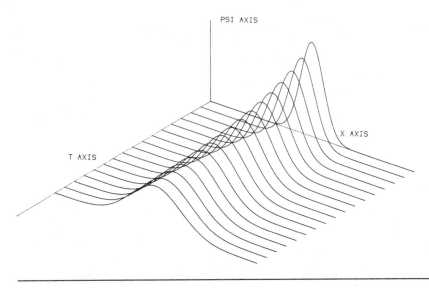

Figure E.3 Space-time plot of an exact solution to the diffusion equation using the procedure "frame" of library "plibd.i", as called in program "diffuse" (Figure E.2).

ENSEMBLE EVOLUTION

Let us assume for ease of discussion that the include file

plibd2.i

contains just the "ensemble" procedure. In a program that includes "plibd2.i" we can start with

```
const dt = 0.001;
     dim = 2;        (*two space dimensions in example*)
     num = 2;        (*there will be two particles this time*)
type ensemble = array[1..dim,1..num] of real;
```

to model a mechanical system having two objects (num = 2) moving on a plane (dim = 2). If the objects experience equal accelerations toward each other, we can calculate trajectories by

```
var position, velocity, acceleration:     (*compute  acceleration[i,j]  for  all
ensemble;                                  four combos of (i,j)*)
advance(velocity,acceleration);            (*this gets new velocity*)
advance(position,velocity);                (*this gets new position*)
```

loop back to get new acceleration components, and so on.

As with most libraries, real usefulness for long projects comes only when we have added powerful application-dependent procedures to programs or to the

```
(* plibd.i  -    dynamical models library *)
(* globals for 'ax' and 'frame' are:

   const max = ...  array to be plotted will have indices -max..max
   type  wave = array[-max..max] of real;
*)
(* globals for 'advance' are:
   const dt =     small time increment
         dim =    spatial dimension of ensembles
         num =  # particles per ensemble
   type ensemble = array[1..dim,1..num] of real;
   *)

procedure ax;
    (* draw axes for x, t, and wave psi *)
    const a=-1;
          b=0.8;
          c=0.85;
    begin
    splot(0.8,-0.3,-1,a,b,c);
    sdraw(-0.8,-0.3,-1,a,b,c);
    sdraw(-0.8,0.3,-1,a,b,c);
        smove(-0.8,-0.3,-1,a,b,c);
    sdraw(-0.8,-0.3,1,a,b,c);
    end;
procedure frame(psi:wave;t:real);
    (* draw one frame of array psi[comp,ii] as ii=1,...,num *)
    const a=0.85;b=0.8;c=-1;
    var ii:integer;
    begin
            for ii:= -max to max do begin
            if ii=-max then smove(-0.8,psi[ii]-0.3,-1+t,a,b,c)
                    else
                    sdraw(0.8*ii/max,psi[ii]-0.3,-1+t,a,b,c);
            end;
    end;
procedure advance(var e:ensemble;de:ensemble);
(* advance the ensemble 'e', having first derivative 'de', by 'dt' *)
    var ii,jj:integer;
    begin
            for jj:= 1 to dim do begin
                for ii:=1 to num do begin
                    e[jj,ii]:= e[jj,ii] + de[jj,ii] * dt;
            end;
            end;
    end;
```

Figure E.4

library itself. One useful modification is to define space-time functions as *ensembles* with declarations such as

```
const dim = 1
      num = 201;    (*say*)
type  ensemble = array[1..dim,1..num] of real;
var psi: ensemble;
```

and then modify the "frame" procedure to be called according to

```
procedure frame(var psi: ensemble; t: real);
            (*modify to plot, for given time t, the array psi[1,jj] as jj = 1..num*)
```

This modification will give equivalent space-time frames but will allow "frame" and "advance" to work in conjunction, with a minimum of declarations ("wave" is no longer declared).

The differences between types "wave" and "ensemble" have been preserved in the library "plibd.i" chiefly because the more illuminating physical modeling examples suggest that there be at least these two separate types in time-evolution studies (Figure E.4). A wave function psi(x,t) can certainly be thought of as an infinite-dimensional vector defined for times t, but the notion of vector can clash with the notion of coordinates such as x itself. For these reasons, it is recommended that you write customized procedures for situations in which an ensemble is to be displayed in space-time format.

Some of the applications of the dynamical modeling procedures in "plibd.i" are described in Chapter 8.

References

Chapter 1

Atkinson, L. (1980). *Pascal Programming,* John Wiley & Sons, New York.

Barron, D. W. (1980). *PASCAL—The Language and its Implementations,* John Wiley & Sons, New York.

Borgerson, M. J. (1982). *A BASIC Programmer's Guide to Pascal,* John Wiley & Sons, New York.

Cooper, D., and M. Clancy (1982). *Oh! Pascal!,* W.W. Norton & Co., New York.

Cooper, J. W. (1981). *Introduction to Pascal for Scientists,* John Wiley & Sons, New York.

Grogono, P. (1980). *Programming in Pascal,* Addison-Wesley Publishing Co., Reading, Mass.

Jensen, K., and N. Wirth (1974). *Pascal User Manual and Report,* Springer-Verlag, New York.

Moore, J. B. (1982). *PASCAL,* Reston Publishing Co., Reston, Va.

Schneider, G. M., and S. C. Bruell (1981). *Advanced Programming and Problem Solving with Pascal,* John Wiley & Sons, New York.

Schneider, G. M., S. W. Weingart, and D. M. Perlman (1978). *An Introduction to Programming and Problem Solving with Pascal,* John Wiley & Sons, New York.

Chapter 2

Abramowitz, M., and I. Stegun (1965). *Handbook of Mathematical Functions,* Dover Publications, New York.

Crandall, R. E. (1978). "On the '3x+1' problem," *Math. Comp.* 32, 144.

Davidson, R. C., and J. B. Marion (1972). *Mathematical Methods for Introductory Physics with Calculus,* W. B. Saunders Co., Philadelphia.

Gradshteyn, I. S., and I. M. Ryzhik (1965). *Table of Integrals, Series, and Products,* Academic Press, New York.

Kovach, L. D. (1982). *Advanced Engineering Mathematics,* Addison-Wesley Publishing Co., Reading, Mass.

Roberts, J. B. (1977). *Elementary Number Theory—A Problem Oriented Approach,* M.I.T. Press, Cambridge, Mass.

Stearns, S. C. (1981). *Pascal Programming for Biology,* 4th ed., Reed College, Portland, Oreg.

Chapter 3

Bell, E. T. (1965). *Men of Mathematics,* Simon and Schuster, New York.

Birkhoff and Rota (1978). *Ordinary Differential Equations,* John Wiley & Sons, New York.

Chapter 4

Eason, Coles, and Gettinby (1980). *Mathematics and Statistics for the Biosciences,* John Wiley & Sons, New York.

Hardy, G. H., and E. M. Wright (1965). *An Introduction to the Theory of Numbers,* Clarendon Press.

Knuth, D. (1969). *The Art of Programming,* Addison-Wesley Publishing Co., Reading, Mass.

Tipler, P. A. (1969). *Modern Physics,* Worth, New York.

Chapter 6

Buhler, J. P., R. E. Crandall, and M. A. Penk (Apr. 1982). "Primes of the form $n! \pm 1$ and $2.3.5 \ldots p \pm 1$," *Math. Comp.*

Cochran, S. (1980). *Digital Analysis of Birdsong: Evidence for Dialects in the Song Sparrow Melospiza Melodia,* Thesis, Reed College, Portland, Oreg.

Cooley, J. W., and J. W. Tukey (Apr. 1965). "An algorithm for machine computation of complex Fourier series," *Math. Comp.*

Crandall, R. E., and M. A. Penk (Jan. 1979). "A search for large twin-prime pairs," *Math. Comp.*

Ralston, A., and P. Rabinowtiz (1978). *A First Course in Numerical Analysis,* McGraw-Hill Book Co., New York.

Whittaker, E. T., and G. N. Watson (1972). *A Course of Modern Analysis,* 4th ed., Cambridge University Press, New York.

Chapter 7

Abdel-Raouf, M. A. (1982). "On the variational methods for bound state and scattering problems," *Phys. Rept.,* 84(No. 3).

Bailar, J. C., et al. (1978). *Chemistry,* Academic Press, New York.

Harriss, D., and F. Rioux (1980). *J. Chem. Ed. 57.*

Mahan, B. (1969). *University Chemistry,* 3rd ed., Addison-Wesley Publishing Co., Reading, Mass.

Chapter 8

Bhatnagar, P. L. (1979). *Non-Linear Waves in One-dimensional Dispersive Systems,* Clarendon Press.

Crandall, R. E. (1978). "On the rings of Saturn," *J. Oregon Acad. Sci.*

Crandall, R. E., and M. H. Reno (1982). "Ground state bounds for potentials x^n," *J. Math. Phys.* 23(No. 1).

Feynman, R. P. (1965). *The Character of Physical Law,* British Broadcasting Co.

Fredrick, L.W., and R. H. Baker (1976). *Astronomy,* 10th ed., D. Van Nostrand Company, New York.

Goldberg et al. (1967). *Am. J. Phys.* 35(No. 3).

Goldstein, H. (1980). *Classical Mechanics,* 2nd ed., Addison Wesley Publishing Co., Reading, Mass.

Lamb, Jr., G. L. (1980). *Elements of Soliton Theory,* John Wiley & Sons, New York.

Litt, B. (1981). Thesis, Reed College, Portland, Oreg.

Merzbacher, E. (1970). *Quantum Mechanics,* John Wiley & Sons, New York.

Tufillaro, N. (1981). Thesis, Reed College, Portland, Oreg.

Chapter 9

Beauchamp, K. G. (1973). *Signal Processing Using Analog and Digital Techniques,* G. Allen and Unwin, Ltd., New York.

Crandall, R. E., and S. C. Stearns (1982). "Variational models of life histories," *Theoret. Popul. Biol.* 21(No. 5).

Roughgarden, J. (1979). *Theory of Population Genetics and Evolutionary Ecology: An Introduction,* Macmillan, New York.

Stanford, A. L. Jr. (1975). *Foundations of Biophysics,* Academic Press, New York.

Stearns, S. C., and R. E. Crandall (1981a). "Quantitative predictions of delayed maturity," *Evolution* 35.

Stearns, S. C., and R. E. Crandall (1981b). Bet-hedging and persistence as adaptations of colonizers, *Proc. 2nd Intl. Conf. Sys. and Evol. Bio.*

Kauffman, S. (1972). *Towards a Theoretical Biology,* Waddington (ed.), Edinburgh University Press.

Index